MICROBIAL PHYSIOLOGY

MICROBIAL PHYSIOLOGY

S MEENA KUMARI

Lecturer
Department of Biotechnology
Dhanalakshmi Srinivasan College of Arts and
Science for Women, Perambalur
Tamilnadu

MJP PUBLISHERS
Chennai 600 005

Cataloguing-in-Publication Data

Meena Kumari, S. (1981–   ).
    Microbial physiology / S. Meena Kumari. –
Chennai : MJP Publishers, 2006.
    xx, 345p. ; 21 cm.
    Includes glossary and index.
    ISBN 81-8094-013-6 (pbk.)
    1. Microorganisms – physiology I. Title
    571.29 – dc22   MEE   MJP 011

ISBN 81-8094-013-6
Copyright © Publishers, 2006
All rights reserved
Printed and bound in India

MJP PUBLISHERS
A unit of Tamilnadu Book House
47, Nallathambi Street
Triplicane, Chennai 600 005

Publisher : J.C. Pillai
Managing Editor : C. Sajeesh Kumar
Project Coordinator : P. Parvath Radha

Edited and Typeset at ▓▓ Editorial Services, Chennai-5
Cover : R. Shankari   CIP : Prof K. Hariharan

FOREWORD

This book *Microbial Physiology* is a maiden attempt by Ms. S. Meena Kumari, Lecturer in biotechnology of Dhanalakshmi Srinivasan College of Arts and Science for Women, Perambalur. She deserves to be congratulated for her brave venture. The book is designed to cover all the topics mentioned in the syllabus presented for B.Sc. Microbiology and Biotechnology programmes of the Bharathidasan University. Besides, important topics included in the syllabi of many Indian universities are covered.

The book discusses the principles and concepts of microbial physiology in a highly accessible and comprehensive manner. Special mention should be made of the lucid explanations given for complicated metabolic processes. Suitable illustrations make the book very impressive. Therefore students are sure to understand the complexities in the physiological processes involved in the microbiological systems. Inclusion of questions at the end of every chapter will be of great help to the students.

I am sure that this will be an indispensable book for the undergraduate students of microbiology, biotechnology, botany and zoology.

Dr. R.M. Meenakshi
Principal
Dhanalakshmi Srinivasan College
of Arts and Science for Women

PREFACE

It has been my indulgent pleasure to write a book on microbial physiology and this book is a small endeavour in this field. I have focused on the basic concepts in the physiology of microorganisms which have been covered in 28 chapters. The chapters have all been simplified to the level of the undergraduate students and will definitely trigger the students' curiosity to learn more. A separate appendix of more than 20 basic laboratory experiments in microbial physiology which would help the students in their practicals has also been provided for easy reference. I do hope that this book would be of great use to teachers and students of microbiology. Constructive comments and criticisms are most welcome.

S Meena Kumari

ACKNOWLEDGEMENTS

I am greatly indebted to so many people who have helped me in every way in fulfilling the task of writing this book.

First and foremost I am extremely grateful to our principal Dr. R.M. Meenakshi for her valuable guidance, encouragement, support and help in a great way.

I wish to express my deep sense of gratitude to our chairman Sri A. Srinivasan, vice chairman Sri. S. Kathiraven, Secretary Sri.P. Neelraj for their support at every stage of this work.

My sincere thanks are due to C. Sajeesh Kumar, Managing Editor and Parvath Radha, Project Coordinator, MJP Publishers for their constant help, encouragement and knowledgeable suggestions.

The immeasurable encouragement that I constantly received from my family members and well-wishers were highly instrumental in invigorating me to complete the book in time.

I am very grateful to my friends Mr. Murugesan and Ms. Vasuhi for their dynamic support and encouragement. They have rendered great help in the fine-tuning of my work.

My sincere thanks are due to my friends E. Yeshima, K. Gandhimathi and M. Kalpana for their continuous inspiration, motivation and never-ending help.

I would like to thank Ms. V. Muthumani, Head, Department of Microbiology, Ms. R. Sasidhara, Head, department of Biotechnology, and Ms. Jeyalakshmi, Head, Department of computer science and staff members of the departments of microbiology, biotechnology and bioinformatics for their constant support.

Last but not the least, I extend my sincere thanks to the lab assistants of Microbiology, Biotechnology and programmers of Bioinformatics for their timely help throughout my work.

S Meena Kumari

CONTENTS

1

AN INTRODUCTION TO MICROBIAL WORLD

Since the dawn of life some 3.5 billion years ago, our planet has collided with giant asteriods, erupted with explosive volcanoes at the rates of 20 Mount Saint Helen-sized blasts per month, accumulated in its atmosphere one of the most lethal chemicals in the history of life, endured three ice ages that extended glaciers well into what today are temperature zones, and other natural disasters have clamied some of the earth's most magnificent species. The giants of the terrestrial earth, the dinosaurs, may have fallen victims to one of these episodes, 65 million years ago in the most famous and perplexing vanishing act of all.

But perhaps the most profound episode of extinction, claiming almost every species on this planet, killed off not the earth's largest organisms, but its smallest. The little-publicized disaster occurred some 2 billion years ago, caused by organisms smaller than the period at the end of the sentence. These organisms produced a deadly gas that accumulated in the atmosphere and poisoned the majority of life forms existing on the planet at that time. The earth has never been the same, for that deadly gas is still with us today. It is oxygen.

At that time, the earth was populated exclusively by microorganisms. Microorganisms or microbes are organisms that are too small to be seen with the unaided eye. Of course many microbes escaped extinction, and their descendants are still with us. Among today's microorganisms are a group of pests we commonly refer to as "germs". Considering our problems with germs, extinction of all the earth's microbes might not seem like such a bad thing. We live in a society in which many people border on having phobias about microorganisms, believing them all to be germs. But suppose some global change did precipitate the selective loss of all earth's microorganisms without directly killing any larger organisms, what would be the effect?

LIFE WITHOUT MICROBES

Many people would celebrate the loss of all microbes as a wonderful event, and at first it might seem to be so. Colds, influenza, AIDS, cholera, tuberculosis, and all other infectious diseases, would immediately vanish. It would no longer be necessary to spend billions of dollars on preservation techniques to prevent the mocrobial spoilage of food or decomposition of useful products. Termites in the home would no longer be a problem, for without wood-digesting microorganisms in their guts these insects would starve.

In fact every food chain on earth would eventually be extinguished. Natural decomposition of dead plants and animals would grind to an immediate halt. The nutrients in the bodies of these dead organisms would not be returned to the soil to be reused by plants. Failing to find sufficient nutrients to grow, plants would disappear and all plant eaters would soon follow. Without plants and animals to eat, humans would quickly be just another of the million of species piling up a thick layer of dead organisms on a dead planet.

Before starving, however, we would have another struggle with breathing, for example without the photosynthetic organisms that live in the ocean and generate about half the world's oxygen, atmospheric oxygen would quickly decline. By then, the loss of microbe-produced supplies such as wine, bread, many medicines, cheese, and soya sauce would probably seem inconsequential. Our sewage facilities would fail to remove enviromental disruptive organic materials from our waste water, but that wouldn't matter any more— we would all be dead.

MICROBIOLOGY—AN ELIXIR OF LIFE

The term Microbiology can be defined in the following ways.

i.   Microbiology, is the study of organisms called microroganisms that are too small to be identified clearly by the unaided eye.

ii. Microbiology in another way, the field of science that studies microorganisms—viruses, archaea, bacteria, fungi, algae and protozoa—is a major scientific discipline that is at the forefront of science leading into the twenty-first century. Although the microbial world is highly ubiquitous and microorganisms were the first living inhabitants of earth, it was not until the seventeenth century that microorganisms were first observed through primitive microscopes. If an object has a diameter of less than 0.1 mm, the eye cannot preceive it at all, and very little detail can be perceived in an object with a diameter of 1 mm. Roughly speaking, therefore, organisms with a diameter of 1 mm or less are microorganisms and fall into the broad domain of microbiology. Microorganisms have a wide taxonomic distribution, they include some metazoan animals, protozoa, many algae and fungi, bacteria and viruses. The existence of this microbial world was unknown until the invention of microscopes, optical instruments that serve to magnify objects so small that they cannot be clearly seen by the unaided human eye. Microscopes, invented at the beginning of the seventeenth century, opened the biological realm of the very small to systematic scientific exploration.

Early microscopes were of two kinds. The first were simple microscopes with a single lens of very short focal length, consequently capable of high magnification; such instruments did not differ in optical principle from ordinary magnifying glasses able to increase an image severalfold, which had been known since antiquity. The second were compound microscopes with a double lens system consisting of an ocular and an objective. The compound microscope, with its greater intrinsic power of magnification, eventually displaced completely the simple instrument; all our contemporary microscopes are of the compound type. However, nearly all the great original microscopic discoveries were made with simple microscopes.

MICROBIOLOGY IN THE TWENTIETH CENTURY

During the last decades of the nineteenth century, microbiology became a solid established discipline with a distinctive set of concepts and techniques both, in large measure, out growths of the work of Pasteur. During the same period a science of general biology also emerged. It was the creation of Charles Darwin, who imposed a new order and coherence in the heretofore anecdotal materials of natural history by interpreting them in terms of the theory of evolution through natural selection. Logically, microbiology should have taken its place, alongside other specialized boilogical disciplines, in the framework of post-Darwinian general biology. In fact, however, this did not occur. For half a century after the death of Pasteur in 1895, microbiology and general biology developed in almost complete independence of one another. The major interests of microbiology in this period were characterisation of agents of infectious disease, the study of immunity and its functions in the prevention and cure of disease, the search for chemotherapeutic agents, and the analysis of the chemical activities of microorganisms. All these problems were both conceptually and experimentally remote from the dominant interest of biology in the early twentieth century; the organization of the cell, its role in reproduction and development and the mechanisms of heredity and evolution in plants and animals. Even the distinctive and original technical innovations of microbiology were of little interest to contemporary biologists; their value became widely recognized only about 1950, when tissue and cell culture began to be applied extensively to plant and animal systems.

However, microbiology did contribute significantly to the developement of the new discipline of biochemistry. The discovery of cell-free alcoholic fermentation by Buchner provided the key to the chemical analysis of energy-yielding metabolic processes. In the first two decades of the twentieth century, parallel studies on the mechanisms of glycolysis by muscle and of alcoholic fermentation by yeast gradually revealed their fundamental similarity. Quite unexpectedty, vertebrate physiologists and microbial biochemists had found a common ground. After years later, the analysis of animal and microbial nutrition revealed another unexpected common denominator—the "vitamins"requried in traces by animals proved chemically indentical with the "growth factors" requried by some bacteria and yeasts. The detailed study of the functions of these substances, conducted for reasons of facility in large measure with microorganisms revealed that they are biosynthetic precursors of a variety of coenzymes, all of which play indispensable roles in the metabolism of the cell. These discoveries, spanning the period from 1920 to 1935, demostrated the fundamental similarities of all living systems at the metabolic level—a doctrine proclamied by biochemists and microbiologists under the slogan "the unity of biochemistry."

The second great advance of biology in the early twentieth century — the creation of the discipline of genetics, formed through the convergence of cytology and mendelian analysis had no immediate impact on microbiology. The first important contact between genetics and microbiology occurred in 1941, when Beadle and Tatum succeeded in isolating a series of biochemical mutants from *Neurospora*. This opened the way to the analysis of the consequence of mutation in biochemical terms, and *Neurospora* joined the fruitfly and the maize plant as a material of choice for genetic research. In 1943, an analysis by Delbruck and Luria of mutation in bacteria provided the technical and conceptual basis for genetic work on these microorganisms. In 1944, the work of Avery, McLeod and McCarty on the process of bacterial genetic transfer known as transformation revealed that it is mediated by free deoxyribonucleic acid (DNA). The chemical nature of the hereditary material was thus discovered. In the mid-1950s microbial biochemistry also set out in new directions. Until that time, it had concerned itself with understanding the chemistry of cellular components and reactions. Then it turned to detailed studies on the coordination and regulation of cellular processes at the levels of enzyme action and controlled expressions of specific genes.

In the mid-1970s a series of discoveries ushered in a new and powerful set of capabilities known as "recombinant DNA technology." Concurrent with the developement of recombinant DNA technology came a major boilogical finding with profound evolutionary significance: the realisation that a group of prokaryotic microorganisms termed archaebacteria are as different from other bacteria as they are from plants and animals.

MICROBIAL PHYSIOLOGY

Microbial physiology is a tremendous discipline encompassing knowledge existed from the study of thousands of different microorganisms. It is of course impossible to convey all that regarding microbial physiology in one book. However one can build a solid foundation using a limited number of organisms to illustrate key concepts of the field. For a cell to grow efficiently, all the basic building blocks and all the macromolecules derived from them have to be produced in the correct proportions. With complex metabolic pathways, it is important to understand the manner in which a microbial cell regulates the production and concentration of each product. Two very common mechanisms of metabolic and gene regulation follows.

1.   Feedback inhibition of enyzme activity (metabolic regulation).

2.   Repression of enyzme synthesis (genetic regulation).

For each metabolic and genetic regulation, an organism requires nutrition. Using nutrients they synthesize their macromolecules like proteins, carbohydrates, lipids, amino acids, nuleic acids, etc. Generally microorganisms uptake their nutrients depending on their environment, nutrient availability and metabolic activity. They extract their cellular material from those nutrients for their growth and survival. This book gives you a basic and in-depth knowledge on growth and metabolism of microorganisms.

STUDY OUTLINE

*   Living things too small to be seen with the unaided eye are called microorganisms.

*   Microorganisms are important in the maintenance of an ecological balance on earth.

*   Some microorganisms live in humans and other animals and are needed to maintain the animal's health.

*   Some microorganisms are used to produce food and chemicals.

*   Some microorganisms cause disease.

CONCEPT CHECK

1.   What are the microbial activities which support human life on earth?

2.   How can one exploit the study of microbiology to improve the quality of human life?

2

PROKARYOTIC CELL STRUCTURE AND FUNCTIONS

INTRODUCTION

The evolutionary history of bacteria extends back to at least 3.5 billion years. It is now generally thought that the first cell that appeared on the earth was a simple bacteria, possibly related to modern forms that live on sulphur compounds in geothermal ocean vents. The fact that bacteria have endured for so long in such a variety of habitats indicates a cellular structure and function that are surprisingly versatile and adaptable. The cellular organization of a prokaryotic cell can be represented as in the following map:

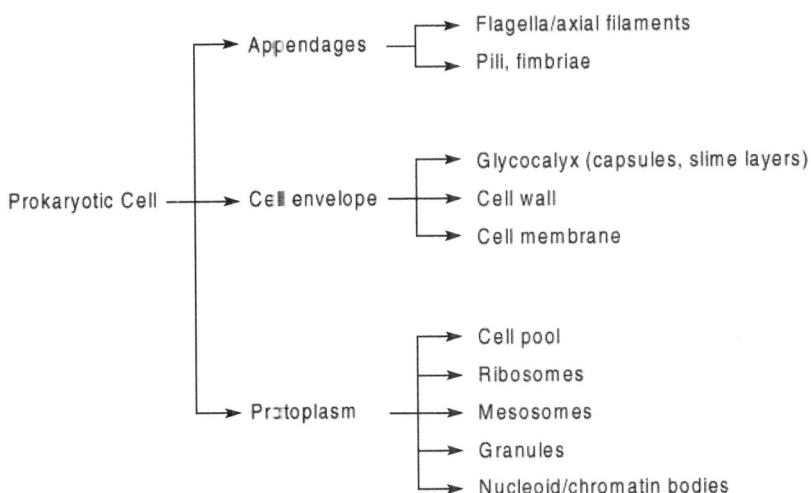

```
                                      ┌──► Flagella/axial filaments
                   ┌──► Appendages ───┤
                   │                  └──► Pili, fimbriae
                   │
                   │                       ┌──► Glycocalyx (capsules, slime layers)
Prokaryotic Cell ──┼──► Cell envelope ─────┼──► Cell wall
                   │                       └──► Cell membrane
                   │
                   │                   ┌──► Cell pool
                   │                   ├──► Ribosomes
                   └──► Protoplasm ────┼──► Mesosomes
                                       ├──► Granules
                                       └──► Nucleoid/chromatin bodies
```

Almost all bacterial cells have a cell membrane, and a central part of cell pool, ribosomes, chromatin bodies; most of the bacteria have cell wall and some have a surface coating or glycocalyx. Some specific structures are also found in some bacteria like flagella, pili, fimbriae, capsules, slime layers, and granules.

OVERVIEW OF PROKARYOTIC CELL STRUCTURE

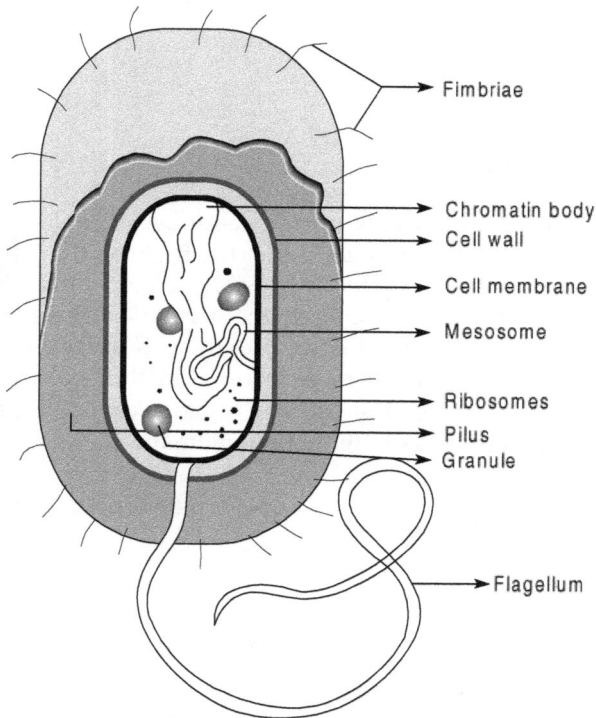

Figure 2.1   Structure of prokaryote.

Bacterial cells, when viewed under the microscope, are featureless and two-dimensional. This chapter reveals a detailed information of each cell component of the prokaryotic cell. The general structure of a prokaryotic cell is shown in figure 2.1. Each cell component has specific function which are listed in table 2.1.

Table 2.1 Functions of different cell components.

Component	Function
Plasma membrane	Selective permeable barrier, mechanical boundary of cell, nutrient and waste transport, location of many metabolic processes (respiration, photosynthesis) detection of environmental factors for chemotaxis.
Gas vacuoles	Buoyancy for floating in aquatic environments.
Ribosome	Protein synthesis.
Inclusion bodies	Storage of carbon, phosphate, and other substances.
Nucleoid	Location of genetic material (DNA).
Periplasmic space	Contains hydrolytic enzymes and binding proteins for nutrient processing and uptake.
Cell wall	Gives bacteria shape, and protection from lysis in dilute solutions.
Capsule and slime layers	Resistance to phagocytosis, adherence to surfaces.
Fimbriae and pili	Attachment to surfaces, bacterial mating.
Flagella	Movement.
Endospore	Survival under harsh environmental conditions.

Morphology deals with the study of the size, shape and rearrangement of microbial cell. These features vary with the species of microorganisms. The structure of microbial cell reveals the composition and topology of chemical constituents building the cell wall and the components outside as well as internal to the cell wall.

SIZE AND SHAPE

The size, shape and arrangement of a microbial cell depends on the species to which it belongs. Bacteria are of about $0.1-60 \times 6$ μm in size. However, there is a variation in the dimension of bacilli ($5 \times 0.4-0.7$ μm), pseudomonads ($0.4-0.7$ μm in diameter, $2-3$ μm length) and micrococci (about 0.5 μm diameter). The smallest (e.g. some members of the genus

Mycoplasma) are about 0.3 μm in diameter, approximately the size of the largest viruses (the pox viruses). Recently there have been reports of even smaller cells. Nanobacteria or ultramicrobacteria appear to range from around 0.2 μm to less than 0.05 μm in diameter.

The rigid cell wall of bacteria contributes to its shape. Generally the bacterial cells are spherical (coccus, plural-cocci means berries), elongated rods (bacillus, plural-bacilli), helical rods (spirillum, plural spirilli), pear-shaped (*Pasteuria*), lobed spheres (*Sulfolobus*), rods with squared ends (*Bacillus anthracis*), rods with helically sculptured surface (*Seliberia*) and of changing shape (pleomorphic), etc.

The unicellular cyanobacterial cells are usually spherical (*Chroococcus, Stenedesmus, Anacystis*), some are elongated and multicellular.

The arrangement of cells are more complex in cocci than bacilli. It depends upon the adherence of cells together after division. Different forms of arrangement are given below.

Coccus Forms

There are several groups of cocci based on the number and arrangement of cells (Figure 2.2).

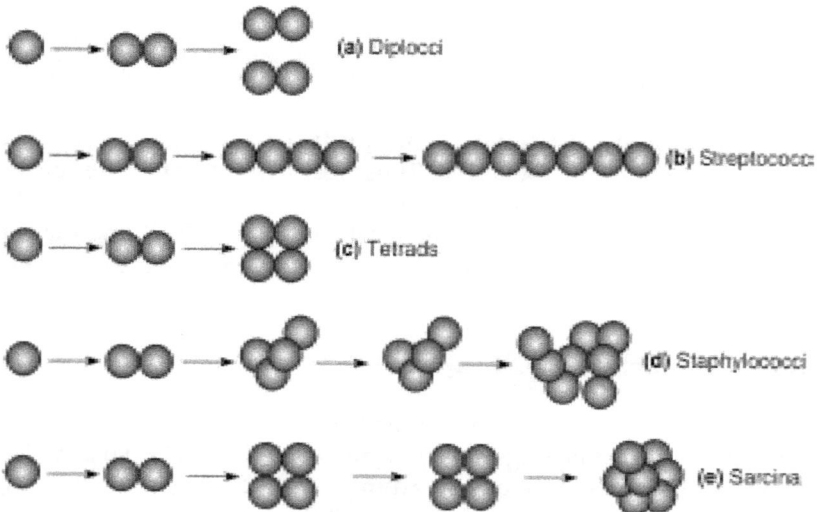

Figure 2.2   Charactersitic arrangements of cocci.

1. *Diplococcus* Cells divide in one plane and get attached permanently in pairs.
2. *Streptococcus* Cells divide in two planes and form groups of four cells.
3. *Staphylococcus* Cells divide in three planes in an irregular pattern producing bunches of cocci.
4. *Sarcinae* Cells divide in three planes in regular pattern producing bunches of cocci.

Forms of Bacilli

There are a few groups of bacilli unlike cocci, as the former divide across their short axes.

1. *Monobacillus* The single elongated cells freely present in nature are monobacillus.
2. *Diplobacillus* After division, the cells remain adhered and appear in paired forms.
3. *Streptobacillus* After division, the cells remain attached in chains appearing like straws.
4. *Coccobacillus* The oval cells looking like cocci are called coccobacilli.

The term bacillus has two meanings, one is the form and the other is the genus. For example, the bacterium *Bacillus anthracis* causes anthrax disease.

Forms of Spirilli

1. *Vibrioid* Bacterial cells having less than one complete twist form vibrioid shape (e.g. *Vibrio cholerae*).
2. *Helical* Cells that have more than one twist form a distinct helical shape, e.g. *Spirillum* (with flagella).

Other Forms

1. *Pleomorphic* These consist of the changing forms e.g. *Rhizobium, Mycoplasma*.
2. *Trichomes* Cells divide in one plane forming a chain which has much larger area of contact between the adjacent cells, e.g. *Beggiatoa, Saprospira*.

3. *Palisade*   The cells are arranged laterally (side by side) to form a matchstick-like structure and at angles to one another, e.g. *Corynebacterium diptheriae.*

4. *Hyphae*   Some microorganisms form multicellular, thin walled, profusely branched filaments called hyphae. The interwoven hyphae are collectively known as mycelium, e.g. *Streptomyces, Aspergillus, Penicillium.*

The cyanobacterial cells are comparatively larger than the bacterial cells. In addition the cells of eukaryotes such as algae, fungi, protozoa, etc. are several times greater in size than the cells of prokaryotes. The unicellular cyanobacteria are usually spherical or elongated. The fungi are either unicellular (yeast, *Candida*) or multicellular hyphal forms (*Fusarium, Aspergillus*).

FLAGELLA

Examination of the bacterial cell reveals various component structures. Some of these are present external to the cell wall like flagella, pili, fimbriae, glycocalyx, slime layer, and capsule. Others are internal to the cell wall. Some structures are present only in certain species. Some are more characteristic of certain species than others; some cellular parts are naturally common to all species. In this topic we will have a clear idea about the external structures of a cell.

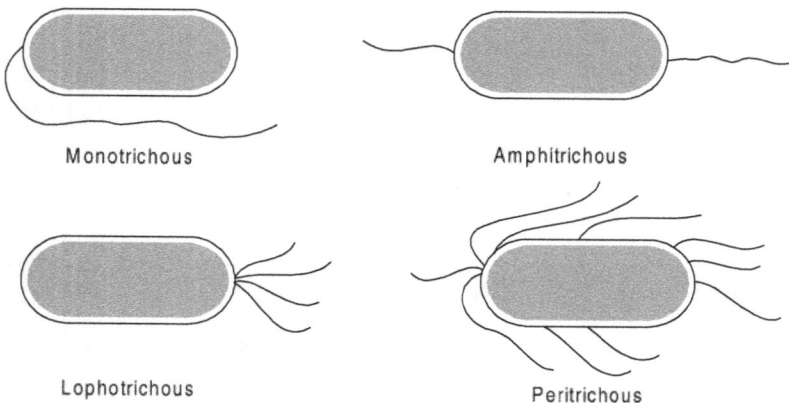

Monotrichous

Amphitrichous

Lophotrichous

Peritrichous

Figure 2.3   Distribution of flagella.

Flagella are hair-like helical appendages (singular, flagellum) that extend through the cell wall and are responsible for swimming motility. They are much thinner than the flagella (or) cilia of eukaryotes, being 0.01–0.02 µm in diameter and they are much simpler in structure. Depending on the species, the flagella varies in number and arrangement. Based on their arrangement, the flagella are classified into four groups.

Monotrichous This type of distribution is present in *Pseudomonas aeruginosa*. There is a single flagellum at one end of the cell. This type of flagella are otherwise called as polar flagella.

Amphitrichous This type of flagella are present in *Aquaspirillum serpens*. This organism has a single flagellum at each pole.

Lophotrichous This type of flagella are found are in *Pseudomonas fluoroscence*. They have two or more flagella at one or both poles of the cell surface.

Peritrichous This type of flagella occur in *E. coli, Salmonella typhi*. In this type, flagella are distributed over the entire cell.

Structure of Flagella

A flagellum is composed of three basic parts. They are the filament, hook and basal body. The filament is the longest and most obvious portion which extends from the cell surface to the tip. The filament is constant in diameter and contains the globular (roughly spherical) protein flagellin arranged in several chains that intertwine and form a helix around a hollow core. The molecular weight of this protein is 30000–60000 dalton. Flagellar proteins serve to identify certain pathogenic bacteria. In most bacteria the filaments are not covered by a membrane or sheath as in eukaryotic cells. The second part of the flagellum is the hook, which is a short, curved segment and links the filament to its basal body which is responsible for the flexibility of the flagellum. The third portion is the basal body which anchors the flagellum to the cell wall and plasma membrane.

The basal body is composed of a small central rod inserted into a series of rings; in *E. coli* and most gram-negative bacteria, the body has four rings connected with the central rod. The outer pair of rings is anchored to various portions of the cell wall like lipopolysaccharide and peptidoglycan. So these rings are named as 'L' and 'P' rings.

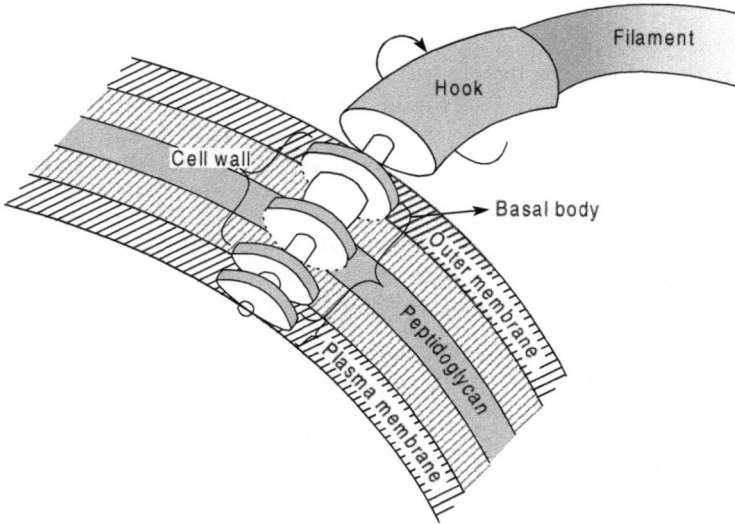

Figure 2.4   Structure of a prokaryotic flagellum. Parts and attachment of a flagellum of a gram-negative bacterium.

The inner pair of rings is anchored to the plasma membrane. So they are called as 'M' ring. In gram-positive bacteria, only the inner pair is present. Also the flagella of eukaryotic cells are more complex than those of prokaryotic cells. In flagellar synthesis there are at least 20–40 genes that are involved. Out of these, ten or more genes code for hook and basal body synthesis. Other genes are concerned with the control of flagellar construction or function.

Regeneration of Flagella

It is believed that flagellin subunits are transported through the filament's hollow, internal core. When they reach the tip, the subunits spontaneously aggregate under the direction of a special filament cap. Thus the filament grows at its tip rather than at the base.

Bacteria with flagella are motile; that is they have the ability to move on their own. Each prokaryotic flagellum is a semi-rigid, helical rotor that moves the cell by rotating from the basal body. The movement of flagellum is either clockwise or anti-clockwise around its long axis (eukaryotic flagella, by contrast, undulate in a wave-like motion). The movement of prokaryotic flagellum results from rotation of its basal body and is similar to the movement of the shaft of an electric motor. As the flagella rotate, they form a bundle that pushes against the surrounding liquid and propels the bacterium. Although the exact mechanochemical basis for the biological "motors" is not completely understood, it depends on the cell's continuous generation of energy.

Types and Mechanism of Flagellar Motility

There are generally two mechanisms involved in the movement of flagella. (a) Long run and (b) twisting or tumbling. These two mechanisms initiate or stop the flagellar movement. The motility of the flagella depends on the clockwise or anticlockwise rotation of the flagellum. If a flagellum moves in the clockwise direction, it simulates the organism to move forward, and is termed as long run. Alternatively if a flagellum rotates in the anticlockwise direction, the movement of the organism stops. This is termed as tumbling or twisting. Movement generally depends on the concentration of the nutrient or repellent in the environment. The flagellar motor can rotate rapidly. For example, *E. coli* flagella can rotate at the speed of 270 rotation/sec.

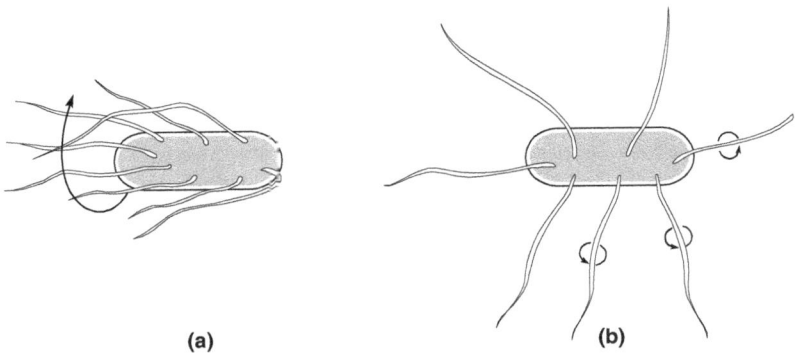

(a) **(b)**

Figure 2.5 Diagrammatic representation of long run (a) and tumbling (b) of flagella.

Generally microorganisms show different types of motility depending on the nature of flagella. They are (i) gliding motility (ii) swimming motility. Gliding motility is found in some bacteria, e.g., *Cytophaga* species. These bacteria are motile only when they are in contact with a solid surface. As they glide they exhibit flexing motion. *Spirochetes* exhibit swimming motility on highly viscous media. They lack the external flagella. Instead of flagella they have flagella-like structures called axial fibrils (or) endoflagella located within the cell just beneath the outer cell envelope. These structures are termed as periplasmic flagella. These axial fibrils are responsible for the motility of spirochetes.

Functions of Flagella

One advantage of motility is that it enables a bacterium to move towards a favourable environment or away from an adverse one. The movement of a bacterium towards or away from a particular stimulus is called taxis. Such stimuli include chemicals (chemotaxis), light (phototaxis) and the earth's magnetic field (magnetotaxis). Motile bacteria contain receptors at various locations, such as in or just under the cell wall. These receptors pick up chemical stimuli such as oxygen, ribose and galactose sugar. In response to the stimuli, information is passed to the flagella.

If the chemotactic signal is positive, called as attractant, the bacteria move towards the stimulus with many runs and few tumbles. If the chemotactic signal is negative, called a repellent, the frequency of the tumble increases as the bacteria move away from the stimulus.

Hence, bacterial chemotaxis are of two types:

i.   Positive chemotaxis
ii. Negative chemotaxis

In positive chemotaxis, the organism swims towards the chemical or nutrient. In the case of negative chemotaxis, the organism swims away from the chemical substance or repellent. There are 20 attractants and 10 repellents that have been identified so for. For example in *E. coli,* the chemoattractant like methyl accepting chemo attractants (maps) attract organic compounds like serine, aspartate, ribose sugar, dipeptides and maltose. If the substance is an attractant the organism gives a long run. But if it is a repellent the organism gives a tumble (negative chemotaxis). The repellents are mostly toxic chemicals and antibiotics.

Phototaxis   Some phototrophic bacteria exhibit positive phototaxis towards increasing light intensities.

Magnetotaxis   This type of taxis is found in *Aquaspirillum magnetotacticum.* This organism exhibits directed swimming in response to the earth's magnetic field or to local magnetic field (magnets placed near the culture). This is attributed to a chain of magnetic inclusions like magnetosomes within the cell which allow the cell to become oriented as a magnetic dipole.

AXIAL FILAMENTS

Spirochaetes are a group of bacteria that have unique structure and motility. One of the best known spirochetes is *Treponema pallidum* the causative agent of syphilis. Another spirochete is *Borrelia burgdoferi*, the causative agent of lyme disease. Spirochetes move by means of axial filaments, bundles of fibrils that arise at the ends of the cell beneath the outer sheath and spiral around the cell. Axial filaments, which are anchored at one end of the spirochete, have a structure similar to that

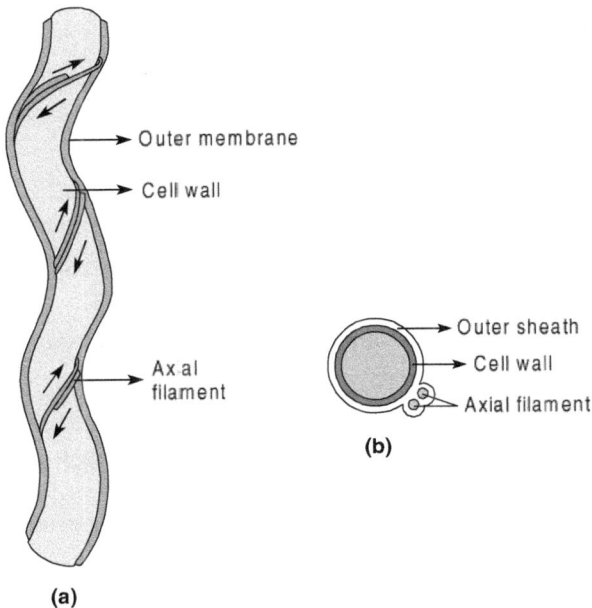

Outer membrane

Cell wall

Axial filament

Outer sheath

Cell wall

Axial filament

(b)

(a)

Figure 2.6 (a) Diagram of axial filaments wrapping around part of a spirochete. (b) Cross section of the spirochete, showing the position of axial filaments.

of flagella. The rotation of the filaments produces a movement of the outer sheath that propels the spirochetes in a spiral motion. This type of movement is similar to the way a corkscrew moves through a cork.

This corkscrew motion probably allows a bacterium such as *T. pallidum* to move effectively through body tissues.

STUDY OUTLINE

* Flagella are relatively long filamentous appendages consisting of a filament, hook and basal body.

* Prokaryotic flagella rotate to push the cell.

* Motile bacteria exhibit taxis, positive taxis is movement towards an attractant, and negative taxis is movement away from repellent.

* Spiral cells that move by means of an axial filament are called spirochetes.

* The axial filament of spirochetes is similar to flagella, except that it wraps around the cell.

CONCEPT CHECK

1. Explain the ultrastructure of flagella.
2. How are flagella arranged in different organisms?
3. How do the flagella activate the organism to run and tumble?
4. Explain the various taxis in flagellar movement.

CRITICAL THINKING

How does an organism recognize repellent and attractant using flagella?

CAPSULES

Some bacterial cells are surrounded by an extracellular polymeric viscous substance forming a covering or envelope around the cell wall. This envelope can be seen under the light microscope after special staining methods and is termed as capsule. The presence of capsule may be detected by negative staining also, using stains such as India ink.

Most bacterial capsules are composed of gelatinous polymers made up of either polysaccharide (*Klebsiella pneumoniae*) or polypeptide *(Bacillus anthracis)* or both. The polysaccharide in the capsule is composed of a single kind of sugar termed homopolysaccharides which are usually synthesized outside the cell from disaccharides by exocellular enzymes. The synthesis of glucan (a polymer of glucose) from sucrose by *S. mutans* is an example. The homopolysaccharide constitutes the capsule of *Acetobacter xylinum*. Heteropolysaccharide capsules are composed of several kinds of sugars that are usually synthesized from sugar precursors that are activated (energized) within the cell, attached to a lipid carrier molecule, transported across the cytoplasmic membrane, and polymerized outside the cell. The capsule generally consists of D-glucose, D-galactose, D-mannose, D-gluconic acid and D-rhamnose, e.g., the capsule of *Klebsiella pneumoniae*. The capsule of pneumococci is made up of hexose, uronic acid and amino sugars and that of streptococci consist of L-amino acids. The bacterial capsule is species-specific and therefore, can be used for immunological differentiation of related species. The amount of these polymers varies with bacterial species. It is sticky in nature and secreted from the inner side of the cell which gets firmly attached to the surface of the cell wall. If the substances are unorganized and loosely attached to the cell wall, the capsules are called slime layers. The freshwater and marine bacteria form Trichomes which are enclosed inside the gelatinous matrix called sheath. Sheath is also found in cyanobacteria and other algae.

A few capsules are polypeptides. For example, the capsule of the anthrax organism, *Bacillus anthracis* is composed entirely of a polymer of glutamic acid. Moreover, this polypeptide is an unusual one because the glutamic acid is the rare D-optical isomer rather than the usual L isomer commonly found in nature.

Functions

Capsules can serve a number of functions, depending on the bacterial species.

1. Capsules may provide protection against temporary drying by binding water molecules.

2. Capsules prevent the attachment of bacteriophages.

3. Capsule protects the cell from phagocytosis, i.e., they are antiphagocytic and inhibit the engulfment of pathogenic bacteria by white blood cells and thus contribute to invasive or infective ability (virulence).

4. The sticky nature of the capsule promotes attachment of bacteria to surfaces. After attachment, they can grow on diverse surfaces, e.g., plant root surfaces, human teeth and tissues (dental carries, respiratory tract), rocks in fast-flowing streams.

5. If capsules are composed of compounds having an electric charge, such as sugar and uronic acid, they may promote the stability of bacterial suspension by preventing the cells from aggregating and settling out because cells bearing similar charged surfaces tend to repel one another.

6. *S. mutans* uses its capsule as a source of energy. It breaks down the sugars of capsules when stored energy is in low amount.

STUDY OUTLINE

•   In many instances capsular material is not highly water-soluble and therefore does not readily diffuse away from the cell that produces it.

•   Capsule appears to be an amorphous gelatinous area surrounding the cell.

•   Electron microscopic observation has revealed that capsules consist of a mesh or network of fine strands.

•   Capsule protects the cells from desiccation as it is hygroscopic and contains water molecules.

CONCEPT CHECK

1. Give an outline about the functions of a capsule.
2. How can we stain the capsule?
3. What are the constituents of a capsule?

CRITICAL THINKING

1. How and why do capsules contribute to the antiphagocytic property of a cell?
2. Which property of the capsule makes a cell virulent?

PILI AND FIMBRIAE

Many gram-negative bacteria contain hair-like appendages that are shorter, straighter, and thinner than flagella and are used for attachment rather than for motility. These structures, which consist of a protein called pilin arranged helically around a central core, are divided into two types, fimbrae and pili, which have very different functions.

Structure of Pili

Pili (singular, pilus) are hollow, nonhelical, filamentous appendages longer than fimbrae and only one or two are present per cell. They are not associated with motility, since they are found on both nonmotile and motile species. Some types of pili play a major role in human infection by allowing pathogenic bacteria to attach to epithelial cells lining the respiratory, intestinal, or genito-urinary tracts. Attachment to these surfaces prevents the bacteria from being washed away by the flow of mucous or body fluids and permits the infection to be established. Pili help bacterial cells in the preparation for the transfer of DNA from one cell to another. According to the function, pili are of two types:

1. Common pili which help to adhere the cell to surfaces and
2. Sex pili which join the other bacterial cells for transfer of genome.

Pili occur either at the poles of bacterial cells or are evenly distributed over the entire surface of the cell. The pili are 0.2–20 µm long with a diameter of about 250A⁰.

Structure of Fimbriae

Fimbriae can occur at the poles of the bacterial cell or are evenly distributed over the entire surface of the cell. Their number ranges up to a few hundred per cell. For example, fimbriae attached to the bacterium *Neisseria gonorrhoea*, the causative agent of gonorrhoea, help the microbe to colonize the mucous membrane. Once colonization occurs, the bacteria can cause disease. When fimbriae are absent (because of genetic mutation), colonization cannot happen and no disease ensues. They originate from the cytoplasm that protrudes outside after penetrating the peptidoglycan layer of the cell wall. Fimbriae are made up of 100% protein called fimbrilin which consists of 163 amino acids. Fimbrilin has a molecular weight of about 16,000 daltons.

Ottow (1975) has classified the fimibrae and pili into the following six groups.

Group 1   Fimibriae of this group act for adherence to a particular surface including the surface of the other cells too. There are about 300 fimibriae/cell arranged peritrichously. Pili of *Neisseria gonorrhoea*, the causative agent of gonorrhoea, help the bacterium to colonize the nuclear membranes. In the absence of pili on cell surfaces, mucous colonization and disease development cannot occur.

Group 2   The sex pili of this group have a uniform diameter of about 9 μm and length of about 1–20 μm. There are about 10 pili/cell. They are filamentous and determined by sex factor. The plasmid carries genes that code for synthesis of sex pili. They make contact between two cells.

Group 3   Fimbriae of this group is peculiar and are found in *Agrobacterium*. They are thick and resemble hollow tubes.

Group 4   This group consists of pili which are flexible, rod-shaped and polar. These are found in the species of *Pseudomonas* and *Vibrio*.

Group 5   Fimbriae of group 5 are polarly arranged and contractile in nature. They are found in *Agrobacterium* spp., *Pseudomonas rhodes* and *Rhizobium lupini*. They contract and bring two bacterial cells to close contact and therefore promote the conjugation process.

Group 6   Group 6 is the characteristic bundles of fimbriae found in gram-positive *Corynebacterium renale*. The filaments function as specific antigens.

Functions of Pili and Fimbriae

1. Bacteria having fimbriae are called fimbriate bacteria. These bacteria show adhesive properties, can attach the organism to the natural substrate or to the other organism. Fimbriae agglutinate the blood cells such as erythrocytes and leucocytes, epithelial cells, etc.

2. Fimbriae contribute the antigenic property as they act as thermolabile nonspecific agglutinogens.

3. Fimbriae also interfere with metabolic activity. The Fm^+ cells (cells containing fimbriae) possess higher rate of metabolic activity than the Fm^- cells (cells devoid of fimbriae). Moreover, they function as aggregation organelles, i.e., they can form stellate aggregation on a static liquid medium.

4. The sex pili make contact between two cells. Since they possess hollow core, they act as conjugaton tubes. The tip of pilus recognizes the female (F^-) cell through which the genetic material of donor (F^+) cell passes to the recipient (female) cell. Only F^- pili (not I pili) contain axial hole (Simon et al., 1978).

STUDY OUTLINE

- Fimbriae and pili are short, thin appendages.

- Fimbriae help cells to adhere to surfaces and for the transfer of DNA from one cell to another.

- Pili and fimbriae are made up of proteins called pilin and fimbrilin respectively.

- The filamentous structure is governed by the sex factor (plasmid) of the bacterium, for example F factor, col I factor and R factor.

CONCEPT CHECK

1. How do pili interfere with the transfer of DNA from one cell to another.

2. Explain the role of pili and fimbriae in microorganisms.

CRITICAL THINKING

In this chapter it is stated that the fimbriae participate in the metabolism of cell. But they are present on the outer surface of the cell. How?

CELL WALL

The cell wall of bacteria is a semi-rigid complex structure, present inside the capsule and which surrounds the plasma membrane. The cell wall gives shape to the cell, and protects the cytoplasmic inclusions. Almost all prokaryotes have cell walls. The rigidity of the cell wall can be readily demonstrated by subjecting bacteria to very high pressure or other severe conditions. Most bacterial cells retain their original shapes during such treatments.

To obtain isolated cell walls for analysis, bacteria usually must be mechanically disintegrated by drastic means, as by sonic or ultrasonic treatment or by exposure to extremely high pressure with subsequent sudden release of pressure. The broken cell wall is then separated from the rest of the composition of distintegrated cells by differential centrifugation. Isolated cell walls devoid of other cellular constituents, retain the original contour of the cells from which they were derived.

Cell walls of bacteria are also used to differentiate the gram-positive and gram-negative bacteria. In prokaryotes, the cell wall of eubacteria is different in its composition from archaebacteria. Also cell walls of eukoryotic microorganisms (e.g. algae, fungi) differ chemically from those of prokaryotes. But in the case of animals, they do not have a cell wall. They only have a plasma membrane. The archaeabacteria are differentiated from eubacteria in their wall thickness rather than chemical composition, which may be the major factor in gram reaction. The walls of gram-negative species are generally thinner (10–15 nm) than those of gram-positive species (20–25 nm). The walls of gram-negative archaebacteria are also thinner than those of gram-positive archebacteria. Gram-negative bacteria consists of two unit membranes of 75A° wide, separated by 100A° wide periplasmic space. Peptidoglycan is present in the periplasmic space in gram-negative bacteria.

Structure of Gram-positive Cell Wall

The gram-positive cell wall consists of a single 20–80 nm thick homogeneous peptidoglycan or murein layer lying outside the plasma membrane. Gram-positive cells are stronger than those of gram negative bacteria. Microbiologists often refer to all the structures present outer to the plasma membrane as the envelope or cell envelope. This includes the cell wall and structures like capsule when present. Gram-positive cells may have periplasm even if they lack a discrete, obvious periplasmic space. The periplasmic space also contains

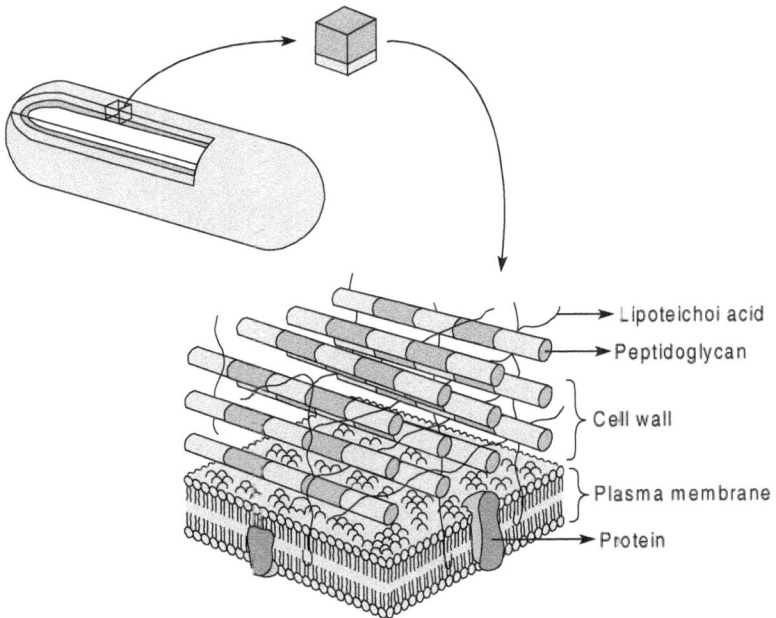

Figure 2.7 Structure of gram-positive cell wall.

enzymes involved in peptidoglycan synthesis and modification of toxic compounds that could harm the cell. In archaebacteria the walls lack peptidoglycan and are composed of protein, glycoproteins or polysaccharides.

Structure of Peptidoglycan

The bacterial cell wall is composed of a macromolecular network called peptidoglycan (murein), which is present either alone or in combination with other substances. Murein is an enormous polymer composed of many identical structures. Peptidoglycan consists of a repeating disaccharide attached by polypeptides to form a lattice that surrounds and protects the entire cell. The disaccharide portion is made up of monosaccharides called N-acetyl glucosamine (NAG) and N-acetyl muramic acid (NAM) (from murus meaning wall) which are related to glucose. N-acetyl glucosamine is connected with different amino acids three of which, D-glutamic acid, D-alanine and meso-diamino pimelic acid, are special types of amino acids which are not found in any other protein. NAM and NAG molecules are linked in rows of 10–65 sugars to form a carbohydrate backbone (the glycan portion of peptidoglycan). Adjacent amino acids are linked by polypeptides (the peptide portion of peptidoglycan) having four alternating D- and

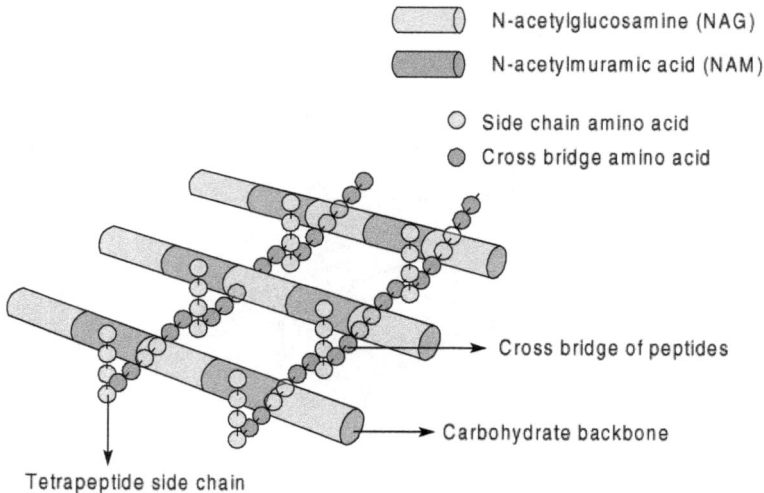

Figure 2.8   Structure of peptidoglycan.

L-amino acids and are connected to the carboxyl group of N-acetyl muramic acid. Many bacteria subtitute another amino acid, usually L-lysine in the third position for meso-diamino pimelic acid. Often the carboxyl group of the terminal D-alanine is connected directly to the amino group of diamino pimelic acid, but a peptide interbridge may be used This peptide interbridge consists of a short chain of amino acids.

In addition, the cell walls of gram-positive bacteria contain teichoic acid, which consists primarily of an alcohol (such as glycerol or ribitol) and phosphate. There are two classes of teichoic acids — lipoteichoic acid, which spans the peptidoglycan layer and is linked to the plasma membrane, and wall teichoic acid, which is linked to the peptidoglycan layer. The presence of phosphate group in the teichoic acid contributes to the negative charge on the gram-positive cell wall. Because of their negative charge, teichoic acids may bind and regulate the movements of cations (positive ions) into and out of the cell. They may also be involved in the role of cell growth, preventing extensive wall breakdown and possible cell lysis (rupturing). Finally, teichoic acids provide much of the wall's antigenic specificity and thus make it possible to identify bacteria by serological means. The cell walls of gram-positive streptococci are covered with various polysaccharides that allow them to be grouped into medically significant types. The cell walls of acid-fast bacteria such as *Mycobacterium*, consist of as much as 60% mycolic acid, a waxy lipid, whereas the rest is peptidoglycan. These bacteria can be stained with the Gram stain and are considered gram-positive.

Structure of Gram-negative Cell Wall

The cell walls of gram-negative bacteria consist of one or few layers of peptidoglycan and an outer membrane. The peptidoglycan is bonded to lipoproteins (lipids covalently linked to proteins) in the outer membrane and is in the periplasmic space, a space between the outer membrane and the plasma membrane. The periplasmic space contains high amount of degradative enzymes like hydrolytic enzymes and transport proteins. They do not contain teichoic acids. Because the wall contains only a small amount of peptidoglycan, they are more susceptible to mechanical breakage. The outer membrane of the gram-negative cell consists of lipoproteins, lipopolysaccharides (LPS) and phospholipids. The outer membrane and peptidoglycan are so firmly linked by the lipoprotein that they can be isolated as one unit. Lipopolysaccharides (LPS) are large, complex molecules that contain both lipid and carbohydrate, and consist of three parts.

1. Lipid A
2. Core polysaccharide
3. O side chain (O antigen)

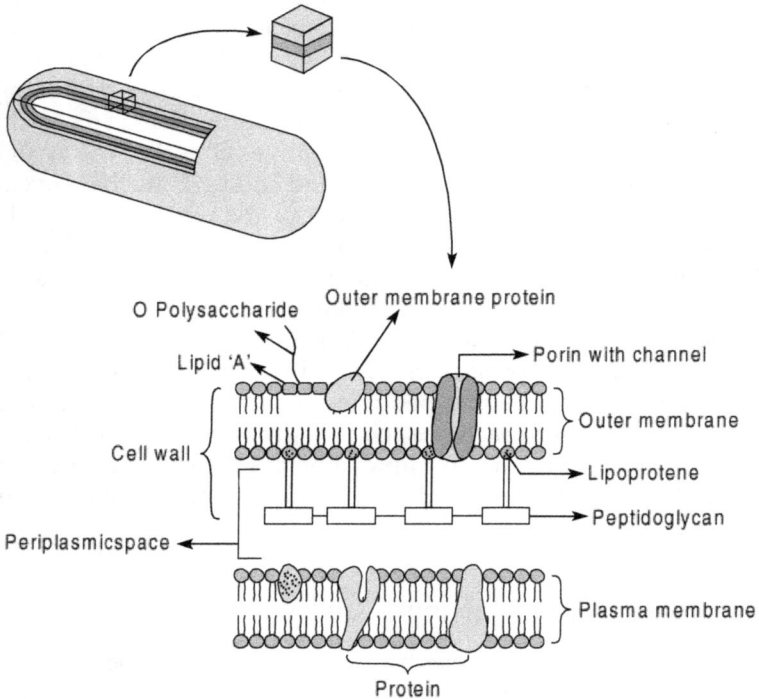

Figure 2.9   Structure of gram-negative cell wall.

The lipid A region contains two glucosamine sugar derivatives, each with three fatty acids and with a phosphate or pyrophosphate attached. This lipid A is buried in the outer membrane and the remainder of the LPS molecule projects from the surface. Core polysaccharide is joined to lipid A. In *Salmonella,* it is constructed of 10 sugars, many of them are unusual in stucture. O side chain, which extends outward from the core, functions as an antigen and is useful for distinguishing species of gram-negative bacteria. For example, over 2000 *Salmonella serovars* (variations within a species) can be distinguished by lipid A which is referred to as endotoxin and is toxic in the host's bloodstream or gastrointestinal tract. It causes fever and shock. Since the core polysaccharide usually contains charged sugars and phosphate group, LPS contributes to the negative charge on the bacterial surface. Lipid A is a major constituent of the outer membrane and the LPS helps to stabilize the membrane structure.

However the outer membrane does not provide a barrier to all substances in the environment because nutrients must pass through to sustain the outer membrane, due to proteins in the membrane called porins, that form channels. Porins permit the passage of molecules such as nucleotides, disaccharides, peptides, amino acids, vitamin B_{12}, and iron. However, they can also make bacteria vulnerable to attack by providing attachment sites for viruses and harmful substances.

Functions of Bacterial Cell Wall

1. Cell wall prevents the cell from rupturing when the osmotic pressure inside the cell is greater than outside the cell.
2. It also helps in maintaining the shape of bacterium and serves as a point of anchorage for flagella.
3. Clinically, the cell wall is important because it contributes to the ability of some species to cause disease and is the site of action of some antibiotics.

STUDY OUTLINE

- Gram-positive cell wall consists of layers of peptidoglycan and also contains teichoic acid.

- Gram-negative bacteria have a lipoprotein–lipopolysaccharide–phospo-lipid outer membrane surrounding a thin peptidoglycan layer.

- The outer membrane protects the cell from phagocytosis and from penicillin, lysozyme and other chemicals.

- Because of its rigid structure, the cell wall gives shape to the cell.

- Its main function is to prevent the cell from expanding and eventually bursting because of uptake of the water, since most bacteria live in hypotonic environments.

CONCEPT CHECK

1. Define, teichoic acid, lipopolysaccharide and porins.
2. How does the cell wall get a negative charge?
3. Explain about the functions of cell wall.
4. Give a detailed account of the structure of gram-postive and gram-negative cell wall?
5. Differentiate betweem gram-positive and gram-negative cell wall.

CRITICAL THINKING

1. In this chapter it is clearly indicated that both gram-negative and gram-positive cell walls have virulence capacity. Which one can act as a better immunogen?
2. Due to the chemical composition of gram-negative cell wall, one cannot use lysozyme to rupture it. In that case, what is your choice of chemical?

A TYPICAL CELL WALL

Some prokaryotes naturally lack their cell wall or have very little wall material. The protoplasts of these organisms is surrounded by only a cytoplasmic membrane, e.g. the members of mollecutes (*Mycoplasma* and *Ureoplasma*). Mycoplasma are the smallest known bacteria that can grow and reproduce outside living host cells. Because of their size and lack of cell wall they can pass through most bacterial filters and were first mistaken for virus. Plasma membrane of this organism has a special type of lipids having sterols which are thought to help and protect them from osmotic lysis. Due to lack of a rigid cell wall the cell can take any shape, viz. cocci, filamentous, discs and rosettes.

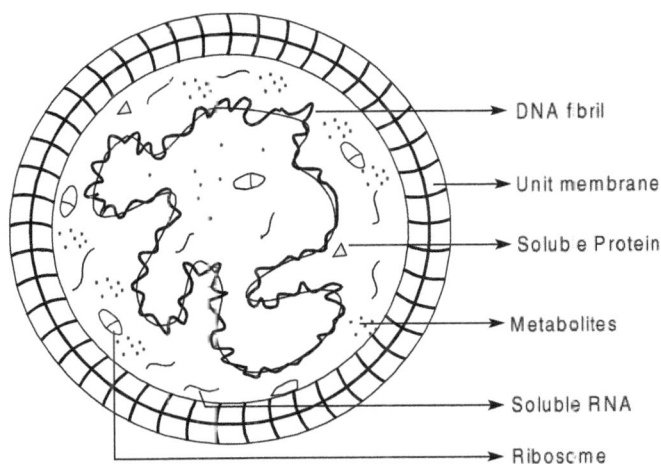

Figure 2.10   A profile of *Mycoplasma* (Diagrammatic).

PLASMA MEMBRANE

Plasma membrane, the layer lying below the cell wall, is otherwise called as cell membrane or cytoplasmic membrane. The term was first coined by C. Nugeli and C. Cramer in 1855. Plasma membrane encloses the cytoplasm of the cell. Cells cannot live without plasma membrane. Prokaryotic plasma membrane consists primarily of phospholipids (the most abundant chemicals in the membrane) and protein. Plasma membrane of eukaryotes contains sterols and carbohydrates. But prokaryotes lack sterol. So they show lesser rigidity than eukaryotes.

Various models have been proposed for plasma membrane so far. But the most acceptable model is the fluid mosaic model which was proposed by Singer and Nicolson in 1972. According to this model, the plasma membrane is a semi-fluid structure in which lipids and proteins are arranged in a mosaic manner. The proteins in this membrane are globular proteins and are of two types: extrinsic (peripheral) proteins and intrinsic (integral) proteins. Extrinsic proteins are soluble in nature and thus dissociate. But intrinsic proteins are water-insoluble and hence cannot dissociate. These proteins are partially embedded in either the outer surface or the inner surface and take part in the lateral diffusion in lipid bilayer. The lipid matrix of membrane has fluidity that permits the membrane components to move laterally. The membrane fluidity is due to the hydrophobic interactions of lipids and proteins. The fluidity is important for a number of membrane functions. Phospholipid and many extrinsic proteins are amphipathic, i.e., they possess both hydrophilic and hydrophobic groups. Phospholipids in plasma membrane are complex in nature and are made up of a glycerol, two fatty acids and, in the place of the third fatty acid, a phosphate group bounded to one or several organic groups.

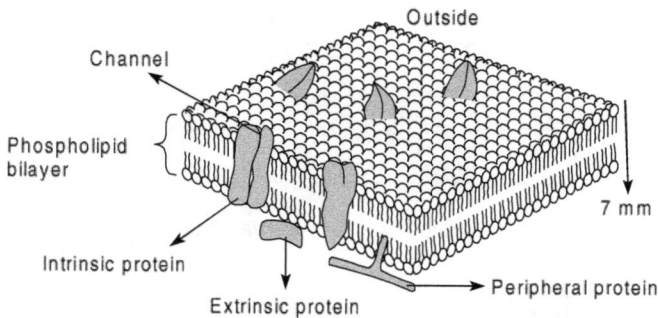

Figure 2.11   Fluid mosaic model of bacterial membrane.

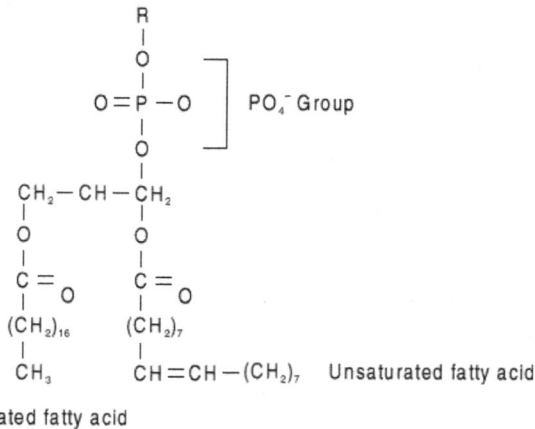

Figure 2.12   Chemical composition of phospholipid molecule.

Phospholipids have polar (hydrophilic) as well as nonpolar (hydrophobic) regions. Polar part contains a phosphate group and a glycerol, while the nonpolar part contains fatty acids. All nonpolar parts of the phospholipids make contact only with the nonpolar portion of the neighbouring molecules. The polar portion occurs towards the outside. This characteristic feature gives the appearance of a bilayer. The semi-fluidity of the plasma membrane is maintained by interspersing unsaturated chains throughout the membrane. Studies have demonstrated that the phospholipid and protein molecules in membranes are not constant, they can move freely within the membrane surface. This mobility is most probably associated with the many functions performed by the plasma membrane. Because the fatty acid tails cling together in the presence of water, phospholipids form a self-sealing bilayer, with the result that breaks and tears in the membrane will heal themselves. They are as viscous as olive oil to allow membrane proteins to move freely enough to perform their functions without destroying the structure of the membrane. The presence of complex lipids becomes a key characteristic of certain microorganisms, e.g. *Mycobacterium* contains high amount of lipid such as waxes and glycolipids which gives the bacterium a distinctive staining characteristic.

In some microorganisms such as mycoplasmas and fungi, sterols are found to be associated with the plasma membrane. These sterols are structurally different from the lipids. Sterols are alcohols composed of hydrocarbon rings attached to a hydrocarbon chain. The sterols separate the fatty acid chains and check packing which harden the plasma membrane at low temperatures.

Functions of Plasma Membrane

1. Plasma membrane serves as a selective barrier through which materials enter and exit the cell. The plasma membrane has selective permeability (sometimes called semipermeability). This indicates that certain molecules and ions pass through the membrane, but others are prevented from passing through it.

2. The organic and inorganic nutrients are transported by permeases through the plasma membrane.

3. It is also important in the breakdown of nutrients and the production of energy. The plasma membranes of bacteria contain enzymes which have the capacity to catalyse the chemical reactions that break down nutrients and produce ATP.

4. It consists of enzymes for biosynthetic pathways that synthesize different components of the cell wall such as peptidoglycan, teichoic acids, polysaccharides, lipopolysaccharides and phosphlipids.

5. It possesses the attachment sites for bacterial chromosome and plasmid DNA.

6. The inner membrane of the plasma membrane invaginates to form mesosomes, a site for respiratory activity. The plasma membrane contains about 200 respiratory proteins that have been found to be anchored for the transport of H^+ ions.

STUDY OUTLINE

• Plasma membrane encloses the cytoplasm and is a phospholipid bilayer with protein (fluid mosaic).

• Mesosomes—irregular infoldings of the plasma membrane—are now considered artifacts.

• Plasma membrane can be destroyed by alcohols and polymyxins.

• The plasma membrane's main function is selective permeability.

CONCEPT CHECK

1. Define the role of plasma membrane in cellular function?

2. Which model of plasma membrane is most acceptable? Explain.

CRITICAL THINKING

In some cases plasma membrane is involved in photosynthesis, i.e., conversion of light energy to chemical energy, for which some special structures are required. How does photosynthesis take place in the plasma membrane?

INCLUSION BODIES

In prokaryotes and eukaryotes the cytoplasm has several kinds of reserve deposits called as inclusion bodies. These inclusion bodies are the bases for identification of certain microorganisms and are common in some microorganisms. But some inclusion bodies are limited to some species. In 1974, Shiveley has given an excellent account of inclusion bodies of prokaryotes. Some of the inclusions are explained below.

Metachromatic Granules

The name is extracted from the staining of these granules to red colour with certain blue dyes like methylene blue. They are collectively known as volutin. Volutin indicates the reserve of inorganic phosphate (polyphosphate) which can be used during ATP synthesis. They are formed in cells that grow in a phosphate-rich environment. Metachromatic granules are found in algae, fungi and protozoa, as well as in bacteria, e.g. *Corynebacterium diptheriae*. Thus they have a diagnostic significance.

Polysaccharide Granules

These granules mainly consist of glycogen and starch, and presence of these granules can be demonstrated with iodine. In the presence of iodine, the glycogen gives reddish brown colour and with starch it gives blue colour.

Lipid Inclusions

Lipid inclusions are found in organisms like *Mycobacterium, Bacillus, Azotobacter* and *Spirillum.* They are unique to bacteria and are mainly composed of a polymer of polyhydroxy-β-butyric acid. These inclusions can be identified by the use of fat-soluble dyes such as Sudan dyes.

Sulphur Granules

Some bacteria like sulphur bacteria obtain their energy by oxidizing sulphur and sulphur-containing compounds. They store the sulphur as sulphur granules inside the cell where they serve as an energy source, e.g. *Thiobacillus*.

Carboxysomes

These inclusions mainly contain the enzyme ribulose 1, 5-diphosphate carboxylase, which is a polyhedral or hexagonal inclusion. This enzyme is mainly required for carbon dioxide fixation for those organisms which use carbon dioxide as their sole source of carbon. These are mainly involved in the photosynthesis process. The bacteria which contain carboxysomes are nitrifying bacteria, cyanobacteria, and thiobacilli.

Gas Vacuoles

Gas vacuoles are mainly found in aquatic prokaryotes including cyanobacteria, anoxygenic photosynthetic bacteria, and halobacteria. They are mainly used for the buoyancy of the organism so that the cells can remain in water at the depth appropriate for them to receive significant amounts of oxygen, light and nutrients. Each vacuole consists of rows of several individual gas vesicles that are hollow cylinders covered by protein.

STUDY OUTLINE

- Inclusions are found in both prokaryotes and eukaryotes.
- In prokaryotes, each inclusion contributes a unique property.
- These inclusions are also used for the identification of the organism.
- Some inclusions like carboxysomes, glycogen granules and metachromatin granules take part in metabolic processes and function as reserve food material.

CONCEPT CHECK

1. Write about the inclusions of prokaryotic cells?
2. Which inclusion in aquatic prokaryotes is responsible for buoyancy?
3. How do you identify the polysaccharide granules using iodine?

CRITICAL THINKING

It has been stated that some bacteria have sulphur granules. But they are considered as minor elements for the cultivation of the microorganism. If so, will its concentration affect the organism?

MESOSOME

Mesosomes are invaginations or localized infoldings of the plasma membrane, and comprise of vesicles and tubules of lamellar whorls. Mesosomes are found only in prokaryotes and not in eukaryotes. Mesosomes are associated with the nuclear area or are found near the site of cell division. The lamellae of mesosomes are formed by flat vesicles arranged parallely. Some of them are connected to the cell membrane. Mesosomes can be observed in *Nitrobacter*, *Nitromonas* and *Nitrococcus*.

Vesicles in the mesosome are formed by invagination and tubular accretion of the plasma membrane. The structure is interrupted because of constrictions at equal distance. This constriction mechanism does not cause the complete separation of tubules. Closely packed spherical vesicles are seen in *Chromatium* and *Rhodospirillum rubrum*. Some scientists suggest that the mesosomes are formed due to vesicularization of the outer half of the lipid bilayer. However, they

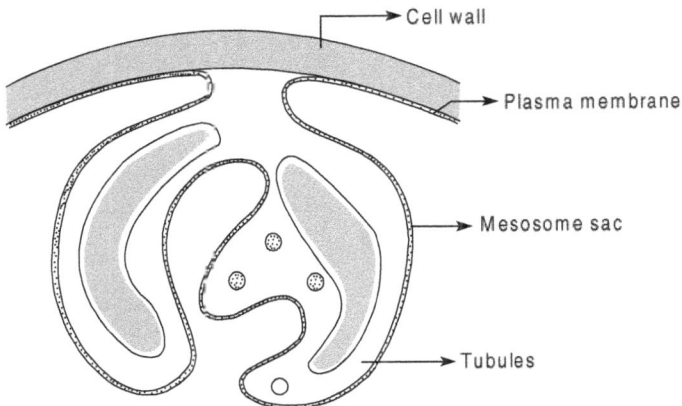

Figure 2.13   Bacterial mesosome.

are the special cell membrane components, the proteins of which differ from that of the cell membrane.

The exact structure and function of mesosomes are not known. They have been suggested to be just artifacts (i.e. a structure that appears in microscopic preparations due to the method of preparation).

Mesosomes also take part in respiration but they are not analogous to mitochondria because they lack outer membrane. Respiratory enzymes have been found to be present in cell membrane. Vesicles of mesosomes have respiratory enzymes, and the components of electron transport such as ATPase, dehydrogenase, cytochrome are either absent or present in small amounts. Mesosomes are believed to be the sites for synthesis of some wall membranes.

Mesosomes might play a role in reproduction also. During binary fission, a cross wall is formed resulting in the formation of two cells. Mesosomes begin the formation of the septum and attach bacterial DNA to the cell membrane. They split the bacterial DNA and provide it to each of the daughter cells. In addition, the foldings of mesosomes increase the surface area of the plasma membrane that in turn increases the absorption of nutrients.

STUDY OUTLINE

- Mesosomes are present in prokaryotes and not in eukaryotes.

- They are formed by the evagination of plasma membrane.

- They contain some enzymes like ATPase, dehydrogenae and cytochrome which take part in the metabolism of the cell.

- They are also involved in the activation of nutrient uptake by the plasma membrane.

CONCEPT CHECK

Describe the structure and functions of the mesosome.

CYTOPLASM

The term cytoplasm of a prokaryote refers to the internal matrix of the cell contained inside the plasma membrane. The water content of the cytoplasm is about 80% and also contains primarily proteins (enzymes), carbohydrates, lipids, inorganic ions, and many low-molecular weight compounds. Cytoplasm is thick and transparent and contains inorganic ions in much higher concentration. Cytoplasm is aqueous, semitransparent and elastic. The important components present in the cytoplasm are DNA, particles called ribosomes, and reserve deposits called inclusions. Eukaryotic cytoplasm differs from prokaryotic cytoplasm in certain features like cytoskeleton and cytoplasmic streaming. Compartmentation of organelles are absent in prokaryotes and present in eukaryotes.

NUCLEAR AREA

Nuclear area in a bacterial cell is otherwise called as nucleoid, which contains a single long circular molecule of double-stranded DNA, the bacterial chromosome. The genetic information of a cell is present in the bacterial chromosome, which carries all the information required for the cell structure and functions. In prokaryotic chromosomes, the histone proteins are absent. But in the case of eukaryotes it is surrounded by the nuclear envelope which gives stability to the chromosome. The shape of the nuclear area is spherical, dumbbell-shaped or elongated. Twenty percent of the cell volume is occupied by DNA in actively growing bacteria because they pre-synthesize nuclear material for future use for their progeny.

The chromosomes are attached with the plasma membrane and they are responsible for the replication of the DNA and segregation of the new chromosomes to daughter cells in division. In addition, the bacteria also contain extrachromosomal DNA which is in the form of small, circular, double-stranded DNA molecules called plasmids. These plasmids are not connected with the bacterial chromosome and they are autonomously replicating genetic elements. Researches indicate that plasmids are associated with plasma membrane proteins. They usually contain from 5 to 100 genes that are generally not crucial for survival of the bacterium. Under normal environmental conditions plasmids may be gained or lost without harming the cell. Plasmids may carry genes for such activities as antibiotic resistance, tolerance to toxic metals, production of toxins, and enzyme synthesis. Plasmids can be transferred from one bacteria to another. Plasmid DNA is used for gene manipulation in biotechnology.

Ribosomes

All eukaryotic and prokaryotic organisms have ribosomes, which are the sites of protein synthesis. Large numbers of ribosomes represent high rate of protein synthesis and vice versa. Cytoplasm of a prokaryotic cell contains about 10,000 ribosomes which accounts up to 30% of total dry weight of the cell. The cytoplasm of a prokaryotic cell contains tens of thousands of these very small structures, which give the cytoplasm a granular appearance.

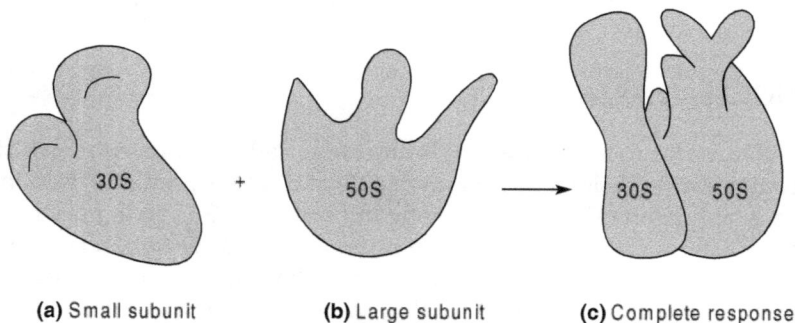

(a) Small subunit   (b) Large subunit   (c) Complete response

Figure 2.14   Structure of Prokaryotic Ribosome.

Ribosomes are composed of two subunits, (Figure 2.14) each subunit being composed of proteins and a type of RNA called ribosomal RNA (rRNA). Ribosomes of prokaryotes are called 70S ribosomes and that of eukaryotes as 80S ribosomes. The letter S refers to Svedberg units, which indicate the relative rate of sedimentation during high-speed ultracentrifugation. Sedimentation rate is a function of the size, weight, and shape of a particle. The 70S ribosome consists of a small 30S subunit containing one molecule of RNA and a larger 50S subunit containing two molecules of rRNA.

Several antibiotics such as streptomycin, neomycin, and tetracyline, work by inhibiting protein synthesis on the ribosomes. Because of differences in prokaryotic and eukaryotic ribosomes, the microbial cell can be killed by the antibiotic while the eukaryotic host cell remains unaffected. The ribosomes of *E. coli* consist of three types of RNA, 5S, 16S and 23S, and 53 proteins; 30S subunit consist of 16S RNA and 21 proteins. The 5S RNA is 120 nucleotides long and 16S RNA is about 1,600 nucleotides long and 23S RNA is 3,200 nucleotides long.

STUDY OUTLINE

* Nuclear material in the cytoplasm is otherwise called as nucleoid. It contains the bacterial chromosome.

* In addition, the nuclear material also contains extrachromosomal DNA, otherwise called as plasmids.

* The ribosomes of prokaryotes are otherwise called as 70S ribosomes and as a result of dissociation they form 50S and 30S subunits.

* Ribosomes are the sites of protein synthesis.

* Prokaryotic ribosomes are most sensitive to antibiotics.

CONCEPT CHECK

1. What is Svedberg unit?
2. Define plasmids.
3. Write briefly about bacterial chromosomes and their role in genetic transformation.
4. Give an account of the prokaryotic ribosome.

CRITICAL THINKING

Plasmids are smaller than bacterial chromosomes. Is it possible to insert a plasmid into a virus in gene technology?

ENDOSPORE

A number of gram-positive bacteria can form a special resistant, dormant structure called "endospore". This is due to depletion of essential nutrients or due to non-avaliability of water. Under such conditions, certain gram-positive bacteria such as those of the genera *Clostridium* and *Bacillus,* form spores internal to the bacterial cell membrane. They are extraordinarily resistant to conditions of environmental stress such as heat, ultraviolet radiation, lack of water, and exposure to many toxic chemicals and gamma radiation. For example the 7500 year-old endospore of *Thermoactinomyces vulgaris* from the freezing muds of Elk lake in Minnesota have germinated when rewarmed and placed in a nutrient medium. The endospores of some gram-positive bacteria cause disease. Spores of *Coxiella burnetii,* which cause Q fever, forms endospore-like structures that resist heat and chemicals and can be stained with malachite green.

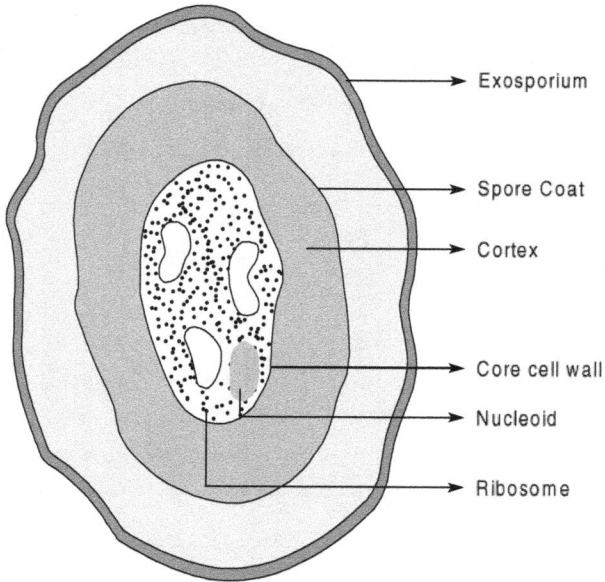

→ Exosporium

→ Spore Coat

→ Cortex

→ Core cell wall

→ Nucleoid

→ Ribosome

Figure 2.15   Structure of Endospore.

Structure of Spore

The spore is often surrounded by a thin delicate covering called exosporium. A spore coat that lies beneath the exosporium is composed of several protein layers and may be fairly thick. It is impermeable and responsible for the resistance of spores to chemicals. The cortex which occupies as much as half the spore volume, rests beneath the spore coat. It is made up of peptidoglycan. The spore cell wall (or core wall) is present inside the cortex and surrounds the protoplast or core. The core has the normal cell structures such as ribosomes and a nucleoid but is metabolically inactive.

The spores are highly resistant to extreme environmental conditions because of the following factors.

1. Presence of dipicolinic acid which is located in the core. It has long been thought that dipicolinic acid is directly involved in spore heat resistance.

2. Some small acid-soluble DNA-binding proteins have been discovered in the endospore. They saturate spore DNA and protect it from heat radiation, desiccation and chemicals.

3. Dehydration of the protoplasts appears to be very important in heat resistance. The cortex may osmotically remove water from the protoplast thereby protecting it from both heat and radiation damage.

4. Finally spores contain some DNA repair enzymes. DNA is repaired during germination.

Sporulation or Sporogenesis

The process of formation of endospore within a vegetative (parent) cell is known as sporulation or sporogenesis. But there is no evidence for the factor which triggers this process. In the first stage of sporulation, the spore septum (ingrowth of the plasma membrane) rotates the newly synthesized bacterial chromosomes and a small portion of cytoplasm. At this time the spore septum becomes a double-layered membrane which surrounds the chromosome and cytoplasm. The completely closed structure is called as forespore. After that, thick layers of peptidoglycan are laid down between the two membrane layers. The next step is the formation of a thick coat which is made up of proteins that lie around the outside membrane. This coat is responsible for the resistance of endospore to many harsh chemicals and extreme environment. The size, i.e., the diameter of the spore is the same as or smaller than that of vegetative cells. The location of the endospore varies depending on the species. Terminal endospores are present at one end, subterminal endospores are present near the terminal end and central endospore lies inside the vegetative cell. While the spore matures, the vegetative cell wall is lysed by the endospore which kills the cell, and thus the endospore is freed. The water present in the forespore cytoplasm is

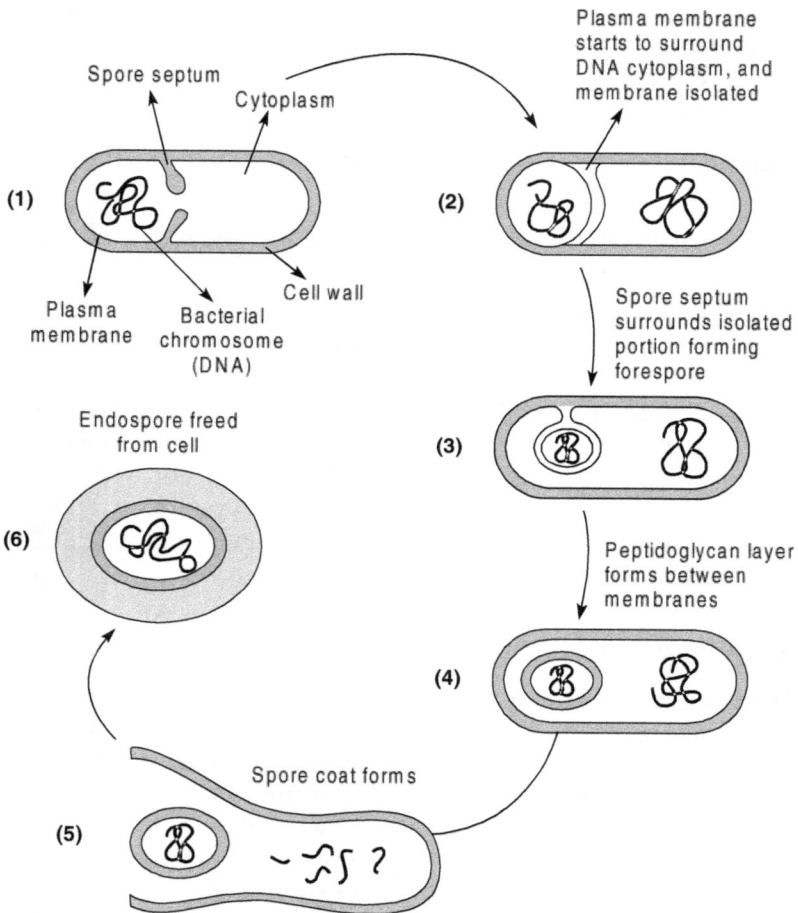

Figure 2.16   Sporulation process of bacteria.

removed by the time sporulation is complete. The highly dehydrated endospore core contains only DNA, small amounts of RNA, ribosomes, enzymes, and a few important small molecules. A spore turns to its vegetative state by a process called germination. It is triggered by chemical damage of the spore coat. The hydrolytic enzyme present in the endospore breaks the surrounding layers of the endospore, water enters and metabolism resumes. One cell of the vegetative form gives rise to a single spore, which after germination remains as one cell. So it means that sporulation is not a method of reproduction of bacteria because it does not increase the number of cells.

STUDY OUTLINE

• Endospores are the resting stage of a vegetative cell.

• They are highly resistant to extreme environmental conditions like, heat, desiccation, radiation and chemicals.

• The presence of dipicclnic acid in the core contributes to the resistivity of the spore.

• Spores can survive up to some thousand years.

CONCEPT CHECK

1. What is a spore? Give some examples for spore-forming bacteria.

2. How are spores classified based on the location? Explain their structure.

3. Explain the process of sporulation.

4. Give details about dipicolinic acid and its heat-resistant property.

CRITICAL THINKING

1. If spores are getting new bacterial chromosomes and cytoplasm, does it mean that they are multiplying?

2. Is it right to use the spore as an indicator for effective sterilization? If yes, why ?

Summarizing the prokaryotic cell structure and functions, we see that although prokaryotes have lesser inner components than eukaryotes, they show some extreme characteristics not seen in eukaryotes. One more important and notable point is that there are no extinct species in prokaryotes like animals and plants. Prokaryotes are ubiquitious in nature. Some harbour our body surfaces and inner parts. They are called as natural flora. They sometimes cause diseases. So they are otherwise termed as opportunistic microorganisms.

PROTOPLAST AND SPHEROPLAST

Bacterial cells have cytoplasmic membrane inside the cell wall which covers the internal organelles. Some chemicals and antibiotics cause damage to the bacterial cell, e.g. antibiotic penicillin and enzyme lysozyme. They cause the damage of the bacterial cell wall. The substances like penicillin and lysozyme have some specificity. Penicillin action is well in the case of gram-negative cell wall. However lysozyme otherwise called as N-acetyl muramidase has high degree of specificity on gram-positive cell wall.

Protoplast

If a gram-positive cell is placed in a solution containing lysozyme or penicillin, denaturation of the cell wall of the organism occurs. The cell membrane with the internal organelles without cell wall is called as "Protoplast". This protoplast is still metabolically active. The lysozyme enzyme catalyses the hydrolysis of the bonds between the sugars in the repeating disaccharide "backbone" of peptidoglycan. Lysozyme digestion is applicable only to gram-positive cell wall, because the gram-positive cells contain high lipid content. These protoplasts remains intact till osmotic lysis does not occur. Because the plasma membrane is semipermeable, if you place the protoplast in an osmotic solution, the cell will easily burst or shrink. Protoplasts are very sensitive to water and dilute salt and sugar concentration. Bacteria normally occur in the hypotonic environment (i.e. environments having a lower osmotic pressure than that within the bacterial cells) and they continually take up water by osmosis; thus they tend to expand, pressing the cytoplasmic membrane against the rigid cell wall. In the absence of a rigid cell wall, there is nothing to prevent the continued expansion and eventual bursting of the protoplast. This bursting can be prevented by preparing protoplasts in an isotonic medium i.e. in a medium that has an osmotic pressure similar to that of the protoplast. Such osmotically protected protoplasts are soft and fragile and are spherical regardless of the original shape of the cell.

Spheroplasts

Lysozyme treatment is also effective in gram-negative bacteria but not like gram-positive cells. In gram-negative bacterial treatment some of the outer membrane also remains. Thus, the cellular contents, plasma membrane, and remaining outer wall layer are called as "spheroplast" also a spherical structure. To proceed the spheroplast formation, the cells are subjected to (Ethlyene diamine tetra acetic acid) EDTA treatment. This treatment will weaken the ionic bonds, and this facilitates the lysozyme to damage the peptidoglycan.

Some antibiotics like penicillin damage the gram-positive bacteria. But in the case of gram-negative bacteria, the lipid layer acts as a barrier. So the gram-negative organisms are resistant to penicillin. But a few gram-negative organisms are susceptible to few B-lactam antibiotics that penetrate the outer membrane better than penicillin.

Some bacteria, like *Mycoplasma*, never have cell walls, and are bounded by only a cytoplasmic membrane; therefore they have many of the properties of protoplasts. Yet they manage to thrive.

Most mycoplasmas are parasites of animals, plants, or arthropods, and therefore live in osmotically favourable or isotonic environments. Some are able to attain a certain degree of rigidity by incorporating cholesterol into their cytoplasmic membranes. Most mycoplasmas have a more or less spherical shape, but *Spiroplasma* consists of helical cells.

STUDY OUTLINE

- Some chemicals and antibiotics cause damage of the bacterial cell wall.

- The plasma membrane with the intact cell internals without cell wall is called protoplast.

- The outer membrane of the cell wall and the cellular content, and plasma membrane are collectively called as spheroplast.

- Both lysozyme and penicillin show high degree of activity on gram-positive cell walls.

CONCEPT CHECK

1. Describe the formation of spheroplast?
2. How does lysozyme treatment mediate the formation of protoplast in gram-positive cell wall.

CRITICAL THINKING

1. One genus of mycoplasma like *Spiroplasma* consists of helical shape. How are such cells able to maintain their shape in the absence of cell wall?
2. What do you mean by 'L' forms?

3

METABOLISM

The term metabolism applies to the assembly of biochemical reactions which are employed by the organism for the synthesis of cell materials and for the utilization of energy from their environments.

The magnitude of metabolism (for example that of bacteria, *E. coli*) can double in number every 40 minutes in a culture medium containing only glucose and inorganic salts, or in 20 minutes in a rich broth. The components of the medium are depleted and very little is added to the medium by the cells. Each cell contains hundreds to thousands of molecules of each of about 2500 different proteins, and about 1000 types of organic compounds and a variety of nucleic acids. It is thus apparent that the bacterial cells participate in a variety of metabolic activities in a remarkable way. The various processes constituting metabolism may be classified into catabolism and anabolism.

CATABOLISM

Those processes which are mainly concerned with the generation of chemical energy in forms suitable for the mechanical and chemical processes of the cells are termed as catabolism (in greek, *cata* = down and *ballein* = to throw).

ANABOLISM

Those processes which utilize the energy generated by catabolism for the biosynthesis of cell components are termed as anabolism (in greek, *ana* = up and *ballein* = to throw). The various activities powered by catabolism include mechanical movement, growth, reproduction, accumulation of foods, elimination of wastes, generation of electricity, maintenance of temperature, etc.

The various anabolic activities may be exemplified by food manufacture, etc.

Classification of General Metabolic Terms

Processes	*Degradative*	*Biosynthetic*
	Catabolism	Anabolism
	Dissimilation	Assimilation
Energy	*Yielding*	*Consuming*
	Exergonic	Endergonic
	Exothermic	Endothermic
Terminal electron	*Oxygenic*	*Not oxygenic*
acceptor	Aerobic	Anaerobic
	Respiration	Fermentation

Metabolism has four specific functions.

1. To obtain chemical energy from the degradation of energy-rich nutrients or from captured solar energy.

2. To convert nutrient material into precursors of cell macromolecules.

3. To assemble these precursors into proteins, lipids, polysaccharides, nucleic acids and other cell components.

4. To form and degrade biomolecules required in specialized functions of cells.

Catabolism and dissimilation are synonyms and refer to the pathways or routes breaking down food material into simpler compounds and resulting in the release of energy contained in them. Anabolism or assimilation defines the utilization of food materials (or intermediates from catabolism) and energy to synthesize cell components.

In dealing with the energy relation of the biological processes, the term exergonic is used to denote a chemical reaction which liberates chemical-free energy. The exothermic process refers to the total energy liberated including heat. The term endergonic denotes the corresponding energy-consuming process.

The term endergonic refers to the processes which require an input of free energy. The word endothermic denotes the total energy requirement including heat.

On the basis of terminal electron acceptor, the term oxygenic refers to the organism which reduces oxygen and hence said to be "aerobe". The route or pathway by which this reaction is accomplished (especially its terminal steps) is called "respiration." The organism which does not reduce oxygen but other compounds is said to be an "anaerobe." However the term fermentation denotes the breakdown of complex molecules into simple compounds with the help of an enzyme. According to modern biochemical view, respiration is defined as the terminal process involved in the reduction of molecular oxygen. The term respiration refers to any process transferring electrons and releasing energy, whether terminating at molecular oxygen or not.

CATABOLIC PATHWAYS (DEGRADATIVE)

There are three stages of catabolism.

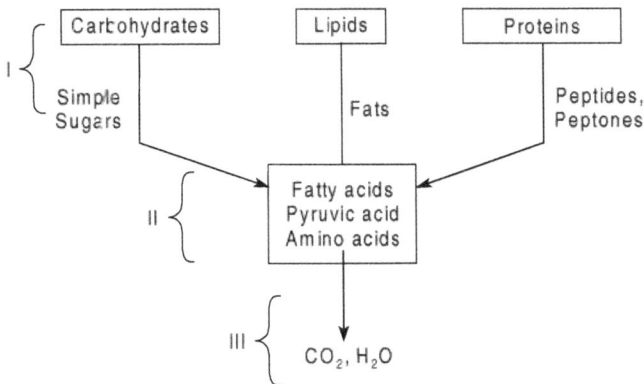

These comprise pathways in which large organic nutrient molecules (carbohydrate, lipids, proteins) are broken down to smaller simpler compounds (e.g. CO_2, NH_2 and lactic acid) frequently involving the participation of oxidation reactions and result in the release of chemical energy contained in the large organic molecules. This energy is then utilized by the organism for growth, movement, replication and also for transduction into other forms of energy such as mechanical, thermal or electrical. Much free energy is conserved in the form of the energy-carrying molecule, adenosine triphosphate (ATP). Some may be conserved as energy-rich hydrogen atoms carried by the coenzyme, NADP, in its reduced form, NADPH.

ANABOLIC PATHWAYS (BIOSYNTHETIC)

These include pathways in which complex organic compounds are produced from simpler precursors, usually involving the participation of reduction reactions, and require an input of chemical free energy which is furnished by the breakdown of ATP to ADP and phosphate. Biosynthesis of some cell components also requires high-energy hydrogen atoms, which are donated by NADPH.

The energy relation between catabolic and anabolic pathways is shown in figure 3.1.

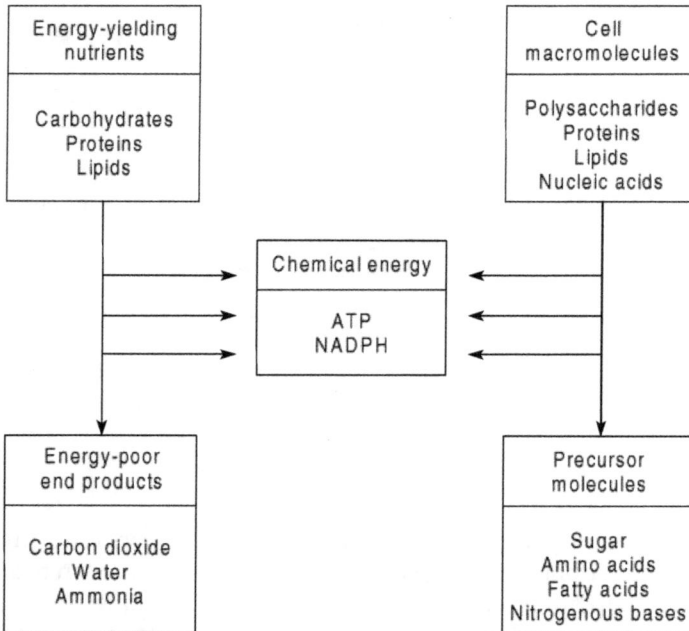

Energy-yielding nutrients	Cell macromolecules
Carbohydrates Proteins Lipids	Polysaccharides Proteins Lipids Nucleic acids

Chemical energy
ATP NADPH

Energy-poor end products	Precursor molecules
Carbon dioxide Water Ammonia	Sugar Amino acids Fatty acids Nitrogenous bases

Figure 3.1   Energy relation between catabolic and anabolic pathways.

ANAPLEROTIC PATHWAY

In this pathway, the product is sometimes synthesized by some subordinate or ancillary routes. Consider the reaction in figure 3.2. The first pathway i.e. from C → D is a normal pathway. But in some way from C → E is an ancillary pathway for product synthesis where each step is mediated by separate enzymes. Such type of reaction is called "anaplerotic pathway".

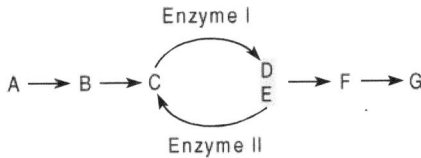

Figure 3.2 Anaplerotic pathway.

Central Metabolic Pathways

Because of their highly diverse nature, bacteria do not have separate metabolic pathways for each of the substances they use. In most cases, they channelize a diversity of materials into relatively few central pathways, pathways that are essential to life and are found in all cellular forms, not just in bacteria. So to a great extent, the ability of bacteria to use a wide diversity of carbon and nitrogen sources for growth reflects their ability to channelize unusual substances, or their metabolics, into central pathways. In this way, the organisms generate energy from a variety of materials through the use of only a few metabolic sequences. Furthermore, the intermediates of catabolism are usually the beginning points for synthesis. It is thus possible for all life forms, through the use of a common collection of central pathways, to both generate energy and synthesize the essential material of life.

The synthesis of a molecule begins at the leave-off point of catabolism. Because of this it is sometimes hard to decide whether a particular metabolic sequence is catabolic or anabolic. Because of this fact, we often regard metabolic pathways as amphibolic. In an amphibolic sequence, the same central pathway participates simultaneously, in both energy-yielding and and energy-requiring processes. Through study of catabolic sequences, we may understand not only the mechanisms by which energy is obtained but also how energy generation and energy utilization are connected.

COMMON BIOSYNTHETIC INTERMEDIATES

In order for an organism to grow, it must synthesize its essential constituents, for example, DNA, RNA, protein, lipids and carbohydrates. Although the cell contains a diversity of materials, they may all be formed from metabolism of twelve compounds, the common biosynthetic intermediates. All life forms, irrespective of the way in which they obtain energy or their nutritional requirements, must synthesize these twelve critical metabolites.

In chemoheterotrophic organisms, those that degrade organic materials and use organic carbon to produce the building blocks for synthesis, it is the task of the organism to provide energy and the critical twelve metabolites. No single metabolic sequence allows the formation of all metabolites but, somehow, they must be formed.

STUDY OUTLINE

- The term metabolism includes both anabolism and catabolism. Metabolism applies to the assembly of biochemical reactions which are employed by the organism for the synthesis of cell material.

- Catabolism is the process which breaks down the complex substances and produces energy during the process.

- Anabolism is the building up process which utilizes the energy for the synthesis of complex materials.

- Some pathways are amphibolic. i.e., the steps are reversible and irreversible.

- In the case of anaplerotic pathways, each step is mediated by separate enzymes.

- Some materials are synthesized by central pathway, i.e., they generate energy and also synthesize the essential material of life.

- For each metabolic reaction, the source and electron donor may be either organic or inorganic substances.

CONCEPT CHECK

1. Explain the metabolism of cell.
2. What are catabolism and anabolism?
3. Define anaplerotic pathway.
4. Write a short note on amphibolic pathway.

CRITICAL THINKING

How do microorganisms survive in nutrient-deficit environment? How do they undergo effective metabolism in such environment?

4

BIOENERGETICS

Energy is defined as the ability to do work. Organisms need energy for their metabolism and biomolecule synthesis. They obtain energy in the form of chemical energy. They obtain energy by the oxidation of substances during their catabolic processes. The energy is conserved in the form of chemical energy as ATP. ATP is the high energy phosphate group compound which is otherwise called as "energy currency". In this chapter, we are going to see how an organism conserves chemical energy.

Chemical energy is the energy released when organic or inorganic compounds are oxidized. In biology, the most commonly used energy units are the kilocalorie (Kcal) and the Kilojoule (KJ). A kilocalorie is defined as the quantity of heat energy necessary to raise the temperature of 1 kilogram of water by 1°C. One kilocalorie is equal to 4.184 KJ. Kilojoule is widely used in microbial energetics. Energy in each chemical can be released in two forms, heat energy and free energy.

FREE ENERGY

Chemical reactions are accompanied by changes in energy. In any chemical reaction some energy is lost as heat. The term free energy is defined as the energy released that is available to do useful work. The change in free energy during a reaction is expressed as $\Delta G^{o\prime}$ where the symbol Δ should read "change in", the supercripts '0' and ''' mean that the free energy value was obtained under "standard" conditions: pH 7, 25°C, at 1M concentration of reactants.

Consider a reaction

$$A + B \rightarrow C + D$$

In the above reaction, the $\Delta G^{\circ\prime}$ is negative, the reaction will proceed with the release of free energy that the cell can conserve in the form of ATP. Such energy-yielding reactions are called "exergonic" reactions. In contrast, if $\Delta G^{\circ\prime}$ is positive, the reaction requires energy in order to proceed; this type of reaction is called as "endergonic". Microbial cells clearly show that endergonic reaction requires energy and exergonic reaction releases energy.

Formation and Calculation of Free Energy $\Delta G^{\circ\prime}$

In addition to knowing the free energy yield of reactions, it is also necessary to know about the free energy of individual substances. This is the free energy of formation, the energy yielded or energy required for the formation of a given molecule from its constituent elements. The free energy formation ($G^{\circ}f$) of the elements (for instance, C, H_2, N_2) is zero. If the formation of a compound from elements proceeds exergonically, then the free energy of formation of the compound is negative (energy is released), whereas if the reaction is endergonic (energy is required), then the free energy of formation of the compound is positive.

Few examples of free energy formation are given below.

Compound	Free energy of formation ($G^{\circ}f$) in KJ/mol
Water (H_2O)	−237.2
Carbon dioxide (CO_2)	−394.4
Hydrogen gas (H_2)	0
Oxygen gas (O_2)	0
Ammonium (NH_4)	−79.4
Nitrous oxide (N_2O)	+104.2
Acetate ($C_2 O_2 O_3$)	−369.4
Glucose ($C_6 H_{12} O_6$)	−917.3
Methane (CH_4)	−50.8
Methanol ($CH_3 OH$)	−175.4

In most compounds the free energy is negative. This tells us that the compounds tend to form spontaneously from elements. The positive $G^{o}f$ for nitrous oxide (+104.2 KJ/mol) tells us that this molecule does not form spontaneously but rather decomposes to nitrogen and oxygen.

It is possible to calculate the change in free energy using free energies of formation. For a simple reaction such as $A + B \rightarrow C + D$, $\Delta G^{o'}$ is calculated by subtracting the sum of the free energies of formation of the reactants.

$$\Delta G^{o'} \text{ of } A + B \rightarrow C + D = G^{o}f\,[C + D] - G^{o}f\,[A + B]$$

Products minus reactants gives the amount of free energy during chemical reactions. However, it is necessary to balance the reaction chemically before free energy calculations can be made.

OXIDATION–REDUCTION

Conservation of energy in living organisms involves oxidation–reduction (also termed redox) reactions. In biochemistry, oxidation and reduction frequently involve the transfer of not just the electron but whole hydrogen atoms. But chemically it implies the addition or removal of electrons. Oxidation denotes the removal of electron or electrons from a substance. Reduction, in contrast denotes the addition of electron or electrons to a substance. A hydrogen atom (H) consists of an electron plus a proton. When the electron is removed, the hydrogen atom becomes a proton (or hydrogen ion, H^+).

Electron Donors and Acceptors

In oxidation–reduction reactions, an electron is being donated by an electron donor and being accepted by an electron acceptor. For example, hydrogen gas, H_2, can release electrons and hydrogen ions (protons) and become oxidized.

$$H_2 \rightarrow 2e^- + 2H^+$$

The above reaction is only a half reaction because electrons do not exist separately in a solution. They need a second reaction. The term half reaction implies the need of a second half reaction. This is because for any oxidation reaction, there should be a subsequent reduction reaction. For example, the oxidation of H_2 could be coupled to the reduction of many different substances including O_2 in a second reaction.

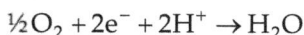

$$\frac{1}{2}O_2 + 2e^- + 2H^+ \rightarrow H_2O$$

This half reaction, which is reduction, when coupled to the oxidation of H_2 above, yields the following overall balanced reaction:

$$H_2 + \frac{1}{2}O_2 \rightarrow H_2O$$

In a reaction of this type, we refer to the substance oxidized, in this case H_2, as the electron donor, and substance reduced, in this case O_2, as the electron acceptor. The key to understand biological oxidations and reductions is to keep straight the proper half reactions—there must always be one reaction involving an electron donor and another reaction involving an electron acceptor.

REDUCTION POTENTIALS

The term reduction potential implies the tendency of a substance to become oxidized or reduced. This potential is measured electrically with reference to a standard substance, H_2 by convention. Reduction potentials are expressed for half reactions and written as reductions. Thus oxidized form is denoted as plus (H^+) and reduced form is denoted as minus (e^-). If protons are involved in the reaction, as is often the case, then the reduction potential is to some extent influenced by the hydrogen ion concentration (pH). By convention in biology, reduction potential is given for neutrality (pH 7) because the cytoplasm of most cells is neutral or nearly so. Using these conventions at pH 7 the reduction potentials (E_o') of

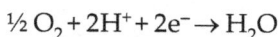

$$\frac{1}{2}O_2 + 2H^+ + 2e^- \rightarrow H_2O$$

is +0.816 V (volts) and that of

$$2H^+ + 2e^- \rightarrow H_2$$

is –0.421 V.

$$H_2 \rightarrow 2e^- + 2H^+$$

The electron-donating half reaction is

$$\frac{1}{2}O_2 + 2e^- \rightarrow O^{2-}$$

The electron-accepting half reaction is

$$2H^+ + O^{2-} \rightarrow H_2O$$

The net reaction for formation of water is

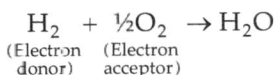

$$\underset{\substack{\text{(Electron} \\ \text{donor)}}}{H_2} + \underset{\substack{\text{(Electron} \\ \text{acceptor)}}}{\tfrac{1}{2}O_2} \rightarrow H_2O$$

ELECTRON CARRIERS

Carriers are molecules which transfer the electrons in an oxidation–reduction reaction from donor to acceptors. They act as intermediates in electron transfer reaction. In electron carrier system, the initial donor is referred to as the primary electron donor and the final acceptor as the terminal electron acceptor. Reduction potentials between the primary donor and the terminal acceptor is used to determine the net energy change. Electron carriers can be classified into two groups based on the nature of the molecule. (i) freely diffusible carriers and (ii) carriers firmly attached to enzymes in the cytoplasmic membrane. These fixed carriers function in membrane-associated electron transport reaction. The diffusible carriers include the coenzyme nicotinamide adenine dinucleotide (NAD^+) and NAD-phosphate ($NADP^+$). NAD^+ and $NADP^+$ are hydrogen atom carriers which always carry 2 hydrogen atoms to the next carrier. This process is termed as dehydrogenation.

The reduction potential of the $NAD^+/NADH$ (or $NADP^+/NADPH$) couple is −0.32 V, which places it fairly high on the electron tower; that is NADH (or NADPH) is a good electron donor. However, although the NAD^+ and $NADP^+$ couples have the same reduction potentials, they generally function in different capacities in the cell. $NAD^+/NADH$ is directly involved in energy-generating (catabolic) reactions, whereas NADP/NADPH is involved primarily in biosynthetic (anabolic) reactions.

Most biological reactions are catalysed by specific enzymes, that can react with only one or a very limited range of substrates. Oxidation–reduction reactions may be considered to proceed in three stages; removal of electrons from the primary donor, transfer of electrons through one or a series of electron carriers, and addition of electrons to the terminal acceptor. However, each step is catalysed by different enzymes each of which binds to its substrate and to its

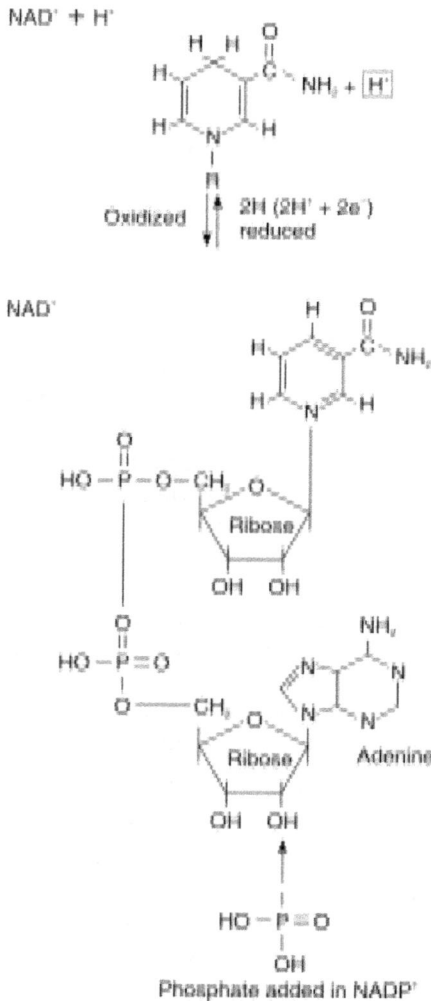

Figure 4.1 Structure of the oxidation–reduction coenzyme nicotinamide adenine dinucleotide (NAD+).

coenzyme specifically. After a coenzyme has performed its chemical function in one reaction, it can diffuse through the cytoplasm until it collides with another enzyme that requires the coenzyme in that form.

HIGH-ENERGY COMPOUNDS AND CONSERVATION OF ENERGY

Energy released in oxidation–reduction reactions is conserved in all living organisms. This energy is used for their cellular functions. In living organisms, chemical energy released in redox reaction is usually conserved in the form of high-energy phosphate bonds; these compounds then function as the energy source to derive energy-requiring reactions in the cell. In phosphorylated compounds, phosphate groups are attached via oxygen atoms by ester or anhydride. However, not all phosphate bonds are high-energy bonds. As a means of expressing the energy of phosphate bonds, the free energy is released when the phosphate bond is hydrolysed. Given below are examples of low-energy and high-energy bonds.

For example

Low-energy ester bond

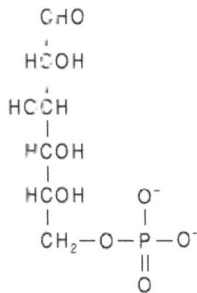

$$
\begin{array}{l}
\text{CHO} \\
\text{HCOH} \\
\text{HCOH} \\
\text{HCOH} \\
\text{HCOH} \quad \text{O}^- \\
\quad\quad\quad | \\
\text{CH}_2-\text{O}-\text{P}-\text{O}^- \\
\quad\quad\quad || \\
\quad\quad\quad \text{O}
\end{array}
$$

Glucose 6-phosphate

High-energy anhydride

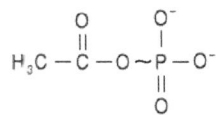

$$
\begin{array}{l}
\quad\quad \text{O} \quad\quad \text{O}^- \\
\quad\quad || \quad\quad\; | \\
\text{H}_3\text{C}-\text{C}-\text{O}\sim\text{P}-\text{O}^- \\
\quad\quad\quad\quad\quad || \\
\quad\quad\quad\quad\quad \text{O}
\end{array}
$$

Acetyl phosphate

In the above example, the $\Delta G^{\circ\prime}$ of hydrolysis of the phosphate bond in glucose 6-phosphate is only –13.8 KJ/mol, whereas the $\Delta G^{\circ\prime}$ of hydrolysis of the phosphate bond in phosphoenol phosphate is –51.6 KJ/mol, almost four times that of glucose 6-phosphate. Thus phosphoenol pyruvate, a phosphoanhydride, is considered a high-energy compound and glucose-6-phosphate, a phosphate ester, is not.

Many cells obtain energy from complex organic molecules like lipid, starch and proteins. During breakdown of these molecules, some of the less energy is released which is used for other reactions. This energy is lost, however, unless it is converted to a usable form that is readily available to the cell whenever it needs energy for life processes. Much of the energy from oxidation reactions is transferred to chemical bonds of high-energy transfer compounds, the most common of which is adenosine triphosphate (ATP).

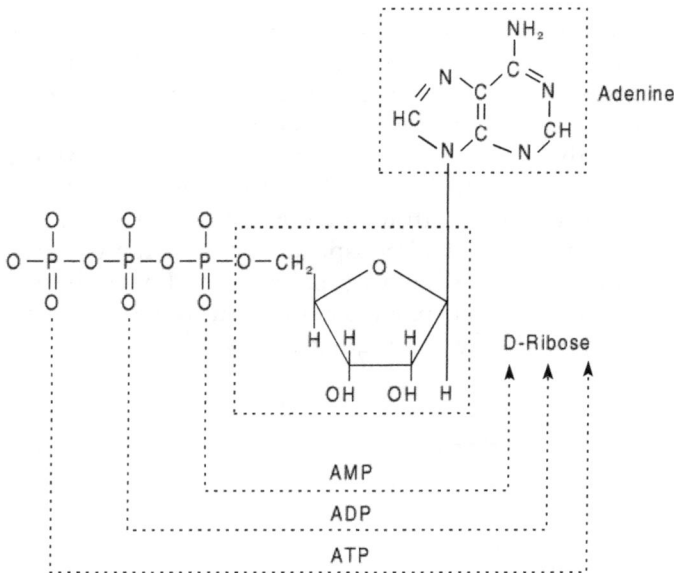

Figure 4.2   ATP—the energy currrency of all organisms.

Adenosine triphosphate contains three phosphate groups attached to each other. A great amount of energy is required to couple adenosine diphosphate (ADP), which has two phosphate with inorganic phosphate (abbreviated Pi) to form a molecule of ATP. This energy is stored in the newly formed "high-energy" phosphate bond. The formation of such high-energy phosphate bond is called phosphorylation. In the reverse reaction (ATP-ADP + Pi + energy), the phosphate bond is broken down (by a process of hydrolysis) releasing the stored energy.

STUDY OUTLINE

- Energy is defined as the ability to do work. Organisms need energy for their metabolism and biosynthesis.

- Organisms obtain energy during oxidation of the catabolic process.

- The term exergonic implies the energy release and endergonic implies the uptake of energy.

- The free energy in each reaction is denoted as $\Delta G^{O\prime}$ where the symbol Δ should be read "change in." The superscripts "O" and "I" mean that the free-energy value was obtained under "standard conditions": pH 7, 25°C, at I M concentration of reactants.

- The energy conserved in living organisms involves oxidation–reduction (also termed redox) reactions.

- The term reduction potential implies the tendency of a substance to become oxidized or reduced.

CONCEPT CHECK

1. Write about a free energy reaction.
2. Define endergonic and exergonic reactions.
3. What are oxidation and reduction potentials?
4. Write about the electron carriers in bioenergetics.
5. Explain about the high-energy phosphate compounds and how energy is conserved in this bond.

CRITICAL THINKING

ATP is considered as a high-energy phosphate bond. But in some reactions, GTP and TTP are involved. These are lower energy compounds than ATP. How do these compounds take part in the metabolic pathway?

5

NUTRITIONAL TYPES OF
MICROORGANISMS

INTRODUCTION

For growth, microorganisms must draw from the environment all the substances they require for the synthesis of their cell materials and for energy generation. These substances are termed nutrients. In other words, nutrients are the substances used in biosynthesis and energy production and therefore are required for microbial growth. Therefore a culture medium must contain, all necessary nutrients in quantities appropriate to specific requirements of the microorganism. However, microorganisms are extraordinarily diverse in their specific physiological properties, and correspondingly in their specific nutrient requirements. Literally thousands of different media have been proposed for their cultivation, and in the description of these media the reason for the presence of the various components are often not clearly stated. The chemical composition of cells, broadly constant throughout the living world, indicate the major material requirements for growth. Water accounts for some 80–90% of the total weight and over 95% of the microbial cell dry weight is made up of major elements or macroelements. The solid matter of cell contains, in addition to hydrogen and oxygen (derivable metabolically from water), carbon, nitrogen, phosphorus and sulphur. All the required metallic elements can be supplied as nutrients in the form of cations of inorganic salts.

Organisms that perform photosynthesis, and bacteria that obtain energy from the oxidation of inorganic compounds typically use the most oxidized form of carbon, CO_2, as the principal source of cellular carbon. The conversion of CO_2 to organic cell constituents is a reductive process which requires a net input of energy. Some organisms utilize oxidation of reduced inorganic carbon sources and some other organisms obtain carbon largely from organic nutrients. The biosynthetic needs of the cell for carbon and organic substrates must often supply the energy requirements of the cell. Consequently, much of the carbon present in the organic substrate enters the pathways of energy-yielding metabolism and is eventually excreted again from the cell as CO_2 or as a mixture of CO_2 and organic compounds.

Organic substrates usually have a dual nutritional role—they serve as a source of carbon and as a source of energy. Many microorganisms can use a single organic compound to supply completely for both their nutritional needs. When organic carbon requirements of individual microorganisms are examined, some show a high degree of versatility, whereas others are extremely specialized. Certain bacteria of the *Pseudomonas* group, for example, can use any one of over 90 different organic compounds as their sole carbon and energy source. Nitrogen and sulphur occur in the organic compounds of the cell principally in reduced form as amino and sulphydryl groups respectively. Some microorganisms are unable to bring about reduction of one or both of these anions and must be supplied with the elements in a reduced form. The requirement for a nitrogen source is relatively common and can be met by the provision of nitrogen as ammonium salts. A requirement for reduced sulphur is rare. It can be met by the provision of sulphur or an organic compound that contains a sulphydryl group (e.g. cysteine).

Some bacteria can also utilize the most abundant natural nitrogen source, N_2. This process of nitrogen assimilation is termed as nitrogen fixation and involves a preliminary reduction of N_2 to ammonia. Some inorganic compounds like potassium (K), calcium (Ca), magnesium (Mg) and iron (Fe^{2+}) can exist in the cell as cations and play a variety of roles. These substances are termed as "minor elements". The term "trace elements" refers to the inorganic elements or a part of enzyme or cofactors and they aid the catalysis of reactions. The trace elements are manganese, zinc, cobalt, molybdenum, nickel and copper. For microorganisms to grow in a medium, the medium requires "grams" of major elements "milligrams" of minor elements and "microgram" of trace elements. Apart from this some microorganisms require some distinct substances to construct their special structure. For example, diatoms need silicic acid (H_4SiO_4) to construct the cell wall. Although most bacteria do not need sodium, those that can grow in saline need sodium ions (Na^{2+}) for their growth.

GROWTH FACTORS

Any organic compound that an organism requires as a precursor or constituent of its organic cell material, but which it cannot synthesize from simpler carbon sources, must be provided as a nutrient. The organic nutrients that are precursors of some components and cannot be synthesized by the organism are called *growth factors*. Most growth factors fall into one of the following three classes in terms of chemical structure and metabolic function.

1. amino acids, required as constituents of proteins.
2. purines and pyrimidines, required as constituents of nucleic acids, and
3. vitamins, a diverse collection of organic compounds that form parts of the prosthetic groups or active centres of certain enzymes.

Only a very small amount of vitamins can sustain growth. Because growth factors fulfil specific needs in biosynthesis, they are required in only small amounts relative to the principal cellular carbon sources, which must serve as the general precursors of cell carbon. Some 20 different amino acids enter into the composition of proteins, so the need for any specific amino acid that the cell is unable to synthesize is obviously not large. The same applies to specific needs for a purine or a pyrimidine; five different compounds of these classes enter into the structure of nucleic acids. The quantitative requirements for vitamins are even smaller, since the various coenzymes of which they are precursors have catalytic roles and consequently are present at levels of a few parts per million in the cell.

Table 5.1 General physiological functions of the principal elements.

Element	Physiological Functions
Hydrogen	Constituent of cellular water, organic cell materials
Oxygen	Constituent of cellular water, organic cell matetrials; as O_2, electror acceptor in respiration of aerobes
Carbon	Constituent of organic cell materials
Nitrogen	Constituent of proteins, nucleic acids, coenzymes
Sulphur	Constituent of proteins (as amino acids cysteine and methionine); of some coenzymes (e.g., CoA, cocarboxylase)

Table 5.1   Contd.

Element	Physiological functions
Potassium	One of the principal inorganic cations in cells, cofactor for some enzymes
Magnesium	Important cellular cation; inorganic cofactor for very many enzymatic reactions, including those involving ATP; functions in binding enzymes to substrates; constituent of chlorophylls
Manganese	Inorganic cofactor for some enzymes, sometimes replacing Mg
Calcium	Important cellular cation; cofactor for some enzymes(for example, proteinases)
Iron	Constituent of cytochromes and other heme or nonhaeme proteins; cofactor for a number of enzymes
Cobalt	Constituent of vitamin B and its coenzyme derivatives
Copper zinc, nickel and molybdenum	Inorganic constituents of special enzymes

Table 5.2   Approximate elementary composition of the microbial cell.

Element	Percentage of dry weight
Carbon	50
Oxygen	20
Nitrogen	14
Hydrogen	8
Phosphorus	3
Sulphur	1
Potassium	1
Sodium	1
Calcium	0.5
Magnesium	0.5
Chlorine	0.5
Iron	0.2
All others	~ 0.3

Based on the requirements of nutrients, microorganisms are classified into the following types. Carbon is required for the skeleton or backbone of all organic molecules and moleculas serving as carbon sources usually contribute both O and H atoms. According to the requirements of carbon, the organisms are classified into two types.

1. By definition, only **autotrophs** can use CO_2 as their sole or principal source of carbon. Many microorganisms are autotrophic and most of these carry out photosynthesis and use light as their energy source. Some autotrophs oxidize inorganic molecules and derive energy from electron transfer, e.g. plants and algae.

2. Some organisms that use reduced preformed organic molecules as carbon source are "heterotrophs" (these preformed molecules normally come from other organisms). Most heterotrophs use reduced organic compounds as sources of both carbon and energy. For example, the glycolytic pathway produces carbon skeleton for use in biosynthesis and also releases energy as ATP and NADH. Laboratory experiments reveal that there are no naturally occurring organic molecules that cannot be used by some microorganisms. Actinomycetes will degrade amyl alcohol, paraffin, and even rubber. For example, *Burkholder capacia* can use over 100 different carbon compounds. *Leptospira* uses only long-chain fatty acids as its major source of carbon and energy.

Based on the source of energy, the organisms can be classified. There are two sources of energy available to organisms (i) light energy and (ii) the energy derived from oxidizing organic or inorganic compounds. According to this *phototrophs* are those organisms which trap light energy as their energy source during photosynthesis. e.g., plants and algae. But the term **chemotrophs** denotes those organisms that obtain energy from the oxidation of chemical compounds (either organic or inorganic).

Based on electron donor, the microorganisms can also be classified. **Lithotrophs** (i.e., rock eaters) use reduced inorganic substances as their electron donor. But in the case of the **organotrophs,** they extract electrons or hydrogen from organic compounds. Despite the great metabolic diversity seen in microorganisms, most may be placed in one of the four nutritional classes based on their primary sources of carbon, energy, and electrons. The majority of microorganisms should be studied as either photolithotrophic autotrophs or chemoorganotrophic heterotrophs.

Photolithotrophic autotrophs, often are called as photoautotrophs or photolithoautotrophs. These microorganisms use light energy and CO_2 for their carbon source. In the case of algae and blue-green bacteria (cyanobacteria), water is employed as the electron donor and hence release oxygen. But in the case of purple and green sulphur bacteria, they cannot oxidize water but extract electrons from inorganic donors like hydrogen sulphide and elemental sulphur. **Chemoorganotrophic heterotrophs,** often called as chemohetero trophs, chemoorganoheterotrophs, or even heterotrophs, use organic compounds as source of energy, hydrogen, electrons, and carbon for biosynthesis. The same organic nutrients satisfy all the requirements of the microorganism. Most of the pathogenic microorganisms belong to this class, e.g. *E. coli.* **Photoorganotrophic heterotrophs,** often called as photoorganoheterotrophs are mostly photosynthetic and they use organic matter as their electron donor and as carbon source. Very few microorganisms belong to this category. This type of requirement is ecologically very important. The photoorganotrophic heterotrophs are common inhabitants of lakes and ponds. **Chemolithotrophic autotrophs** often called as chemolithotrophs, oxidize reduced inorganic compounds such as iron, nitrogen or sulphur molecules to derive both energy and electrons for biosynthesis. These type of microorganisms contribute greatly to the chemical transformation of elements (e.g. the conversion of ammonia to nitrate or sulphur to sulphate).

Some microorganisms have great metabolic flexibility and alter their metabolic patterns in response to environmental changes, e.g. *Beggiatoa*. This organism relies on inorganic energy sources and organic (or sometimes CO_2) carbon sources. These microbes are called as **mixotrophs,** because they combine chemolithoautotrophic and heterotrophic metabolic processes. This will be a definite advantage to the possessor if environmental conditions frequently change.

STUDY OUTLINE

- Nutrients are the substances used in the biosynthesis and energy production and these are required for microbial growth.

- Over 95% of the microbial cell dry weight is made up of major elements (or) macroelements.

- The solid matter of cells contains, in addition to hydrogen, nitrogen, phosphorus and sulphur.

- Some inorganic compounds like potassium (K), calcium (Ca), magnesium (Mg) and iron (Fe^{2+}) can exist in the cell as cations and are termed as "minor elements".

- The trace elements are the inorganic elements or a part of enzyme or cofactors that aid the catalyses of reactions.

- For growth of microorganisms, the media requires grams of major elements, milligrams of minor elements and micrograms of trace elements.

- In addition, the organism requires growth factors like amino acids, vitamins, and purines and pyrimidines.

- Growth factors are the precursors of some components and cannot be synthesized by the organisms.

- Based on the growth requirements, the organisms are classified into various types.

- Autotrophs and heterotrophs are the classification of microorganisms based on their carbon source

- Based on where they derive energy from, the microorganisms are classified into phototrophs and chemotrophs.

- Based on the need of electron donor, microorganisms are typed into lithotrophs (i.e. rock eaters,) and organotrophs.

- Some microorganisms have great metabolic flexibility and alter their metabolic patterns in response to environmental changes. These types of microorganisms are called as mixotrophs.

CONCEPT CHECK

1. Based on the carbon, energy and electron donor requirements, how are microorganisms classified?
2. What are macroelements, microelements and trace elements? How do they function in synthesis of cell constituents?
3. Define mixotrophs.
4. What is growth factor? Explain the role of various growth factors in cellular functions.

CRITICAL THINKING

Microorganisms show great diversity in their habitat. If all the microorganisms are present in one environment, how will they satisfy their needs?

6

MICROBIAL GROWTH

INTRODUCTION

Growth can be defined as the final expression of the physiology of an organism. In the case of microorganisms, it is a very essential event for their survival. The growth of a microorganism reflects the operation of all the structures of the organism in a co-ordinated manner, so that life is possible. The study of growth in bacteria allows an understanding of the critical factors involved in the growth and metabolism of all cells and the spectrum of responses that cells may exhibit in the face of environmental change.

NATURE OF MICROBIAL GROWTH

Growth of a microorganism implies an increase in the amount of protoplasm, the formation of new structures, and eventually, the formation of new cells. The term "balanced" growth means that the cell components increase in proportion to each other and "unbalanced" growth means that components of a new cell increase in a nonconstant relationship to each other. The synthetic activities of microbes that ultimately result in cells may be considered from two perspectives, — how much growth has occurred, and how far it has been accomplished. "Yield" is a measure of the amount of growth obtained and "rate" is a measure of its speed. Both measures are important to the understanding of the factors that are critical to the growth phases.

So growth is a culmination of an ordered interplay of the physiological activities of the cell. It is a complex process involving

1.   Entrance of basic nutrients into the cell.

2. Conversion of their compounds into energy and vital cell constituents.

3. Replication of their genetic material.

4. Increase in size and mass of the cell.

5. Division of the cell into two daughter cells, each containing a copy of the genome and other vital compounds.

Why a Cell Needs Division

Before searching the answer for this question, one should clearly understand the following facts:

1. The relationship of cell size to the division event.

2. The manner in which the components of the cell are replicated.

3. The ways in which the replication of critical cell components are interrelated and co-ordinated.

4. The manner in which DNA replication is related to other cellular processes.

5. The minimum components required for a functional cell.

Clear answers regarding these questions are not entirely apparent at present. However, partial answers and reasonable suggestions are available for all of them. Some of these questions will be considered in this chapter while others will be dealt in later chapters. As regards the cell size, the importance of surface-to-volume ratio appears critical. When an organism decreases in size its surface-to-volume ratio increases, and conversely, when cell size increases, the surface-to-volume ratio decreases. If the cell size is small, the metabolism takes place rapidly and the end products are easily removed. But if the size of the cell increases, the surface-to-volume ratio declines and the rate of metabolism is reduced and the removal of the end product also decreases. Many scientists suggest that the obligatory requirement for cell division in microbes is a reflection of the fact that beyond a certain size, cellular metabolism is insufficient for life and the lessened rate of waste product removal produces a toxic intracellular environment. The cell avoids these difficulties by dividing, thereby restoring a size and surface-to-volume ratio compatible with normal cell process.

A second possible suggestion for the strict requirement for division is the concept of "genetic control". According to this idea, certain amount of genetic material is required to control the given amount of protoplasm. The amount of DNA generally parallels the complexity of an organism. For example, the amount of DNA in bacteria is greater than that in viruses and the DNA in eukaryotic cells is substantially more abundant than that in prokaryotic cells. So, as the complexity of life increases, the amount of DNA required also increases. So genetic control plays a major role in the growth of the organisms. The cell increases to a particular size, switches on the genetic system to control the metabolism, nutrient uptake, replication and other cellular events. When such a point is reached, the cell responds by dividing, restoring an appropriate relationship between the genetic material and the protoplasm.

A third suggestion for the reason a cell divides reflects a consideration of the minimum requirements for the existence of individual functioning cell. For each cell, particular events and components should exist to form a new cell. Each particular cell must contain a genome, a sufficient collection of ribosomes, functioning cytoplasmic membrane, a certain number of proteins, and other entities that allow it to exist as a new cell. According to this concept, division occurs when sufficient synthetic activities have occurred in growing cells to allow a new cell to exist as a separate entity.

Events of Cell Division in a Population

In many types of microbes, particularly bacteria, growth is logarithmic, so during the period of most rapid growth, growth is normally a multiplicative function of time. Because one cell gives rise to two cells and so on, during the most rapid period of growth a plot of the logarithm of numbers as a function of time is linear with a positive slope.

Because of many factors, the rate of viable cell production becomes equal to the rate of death. At this point, no net increase in viable cells occurs and mathematically, a plot of the logarithm of number versus time is a straight line with zero slope. After a further period of time, fewer viable cells are produced than the number that die. As a result, the net number of viable cells decreases. With further time, the relative rates of viable and nonviable cells formed become constant with respect to each other. As a result, net cell death occurs at a constant and logarithmic rate and therefore, a constant fraction of cells die per unit time.

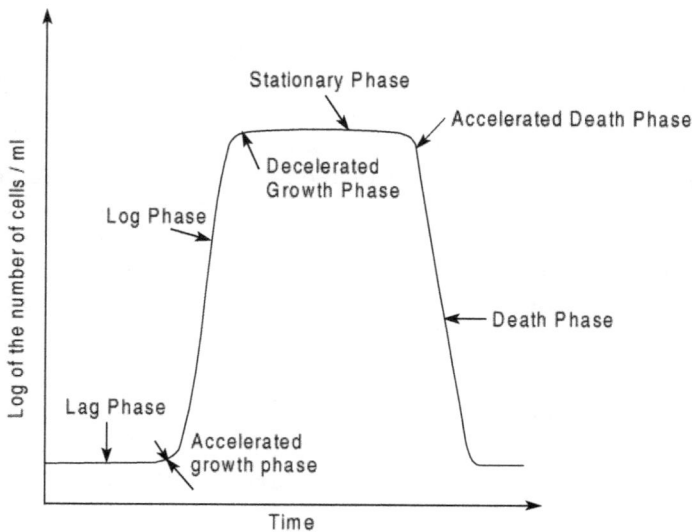

Figure 6.1   Growth stages of batch culture.

GROWTH CYCLES

Under appropriate circumstances, cell division starts immediately and proceeds in an unhampered fashion for a protracted period of time. Prokaryotic cell division follows a geometric progression.

$$2^0 \longrightarrow 2^1 \longrightarrow 2^2 \longrightarrow 2^3 \longrightarrow 2^4 \longrightarrow$$

$$2^5 \longrightarrow 2^6 \longrightarrow 2^7 \longrightarrow 2^8 \longrightarrow 2^9 \longrightarrow \text{etc.}$$

As shown above the number of cells (b) present at a given time may be expressed as

$$b = 1 \times 2^n$$

The total number of cells b is dependent on the number of generations (n = number of divisions) occurring during a given time period. Starting with an inoculum containing more than one cell, the number of cells in the population can be expressed as

$$b = a \times 2^n$$

In this, a is the number of organisms present in the original inoculum. Since the number of organisms present in the population (b)

is a function of the number 2, it becomes convenient to plot the logarithm values rather than the actual numbers. As shown in the growth curve figure, plotting the logarithm of the number, a linear function is obtained. For convenience, logarithm to the base 10 are used. This is possible because the logarithm to the base of a number is equal to 0.3010 times the logarithm to the base 2 of a number. Till this it is clearly indicated that the individual generation time (i.e. the time required for a single cell to divide) is the same for all cells in the population. This can be otherwise termed as doubling time. Now we will see about various phases of growth in a generation. There are four phases. (i) Lag phase (ii) Log phase (iii) Stationary phase (iv) and Death phase.

Lag Phase

This phase is otherwise called as adaptation phase in which division and metabolism take place slowly, in order to synthesize the enzymes and other cell constituents for the uptake of new medium. Actually some of the cells may die in the initial inoculum and so there may be a drop in the number of viable cells. The surviving cells eventually adjust to the new environment and begin to divide at a more rapid rate. If an exponential phase of inoculum is added to the new medium, it shows exponential growth because the cells are actively growing cells and so a new medium also provides a rich environment for their active growth. In a new medium the metabolic rate is also high.

If a stationary phase culture is inoculated in the medium, it shows the lag phase growth. This is because the cells of the stationary phase culture are metabolically dull cells. So the cells are depleted in cell constituents and enzyme. While inoculating in new media, these cells have to synthesize the cell constituents, so it will take time. The lag phase is influenced in many ways. It is a period of intense metabolic activity during which the cells adjust to their new environment. The duration of lag phase depends upon many factors including the present and previous growth conditions. If present conditions are closely similar to previous conditions, the lag phase may be unnoticed. If substantial differences exist between present and previous growth conditions, the lag phase may be extended. If the present environment is more nutritionally abundant than the previous one, the change is known as "shift up" change. Conversely, if the present environment is nutritionally less abundant than the previous one, the change is referred to as a "shift down" change. The period of adjustment required by microbes upon introduction into a new culture system is influenced by many factors other than nutrition. Changes in temperature, pH and gas phase are often influential, since all of these changes require adjustments on the part of the organism. The size of the inoculum substantially influences the lag phase. In general, if other factors are equal, inoculum size and lag time are inversely related.

Log phase

In this phase, the organism is supplied with abundant nutrient and it can grow as rapidly as possible, limited only by its own genetic potential. The metabolic systems of the organism are operating efficiently at their maximum rate. The cells generally synthesize their primary metabolites like acids, alcohols, sugars, etc. This phase is otherwise called as exponential or idiophase. Because of their intense metabolic capacity, logarithmic phase cells are normally more sensitive to physiological change than cells from other growth phases. The growth rate remains constant until conditions in the medium begin to deteriorate (e.g., nutrients are exhausted), since plotting the cell number logarithm during this period results in a linear function as shown in Figure 6.1.

Stationary Phase

All cultures of microorganisms eventually reach a maximum population density at the stationary phase. The precise reasons for entrance into stationary phase are many and are not entirely understood. In the logarithmic phase, the supply of nutrients is abundant and little toxic end product accumulation has occurred, but by the time of the stationary phase, one or more critical nutrients are diminished or exhausted and toxic products, chiefly acids and alcohols are accumulated. Growth of the organism does not take place. Few organisms can grow through metabolism and other bioenergetic ways. Some organisms show slow growth in this phase and some die. The accumulation of acids and alcohols inhibits the growth of the organisms. Logarithmic growth therefore ceases, and the population shifts to a survival mode. Death and growth are balanced events in this phase which is otherwise called as "cryptic growth".

Some organisms need some genes for their survival in this phase, e.g., *E.coli* "Sur" (Survival) genes. Mutation in this gene leads to the sudden death of the organism in the stationary phase. The secondary metabolites that are synthesized by the microorganism in this phase are pigments, toxins, and antibiotics. The entry of microorganisms into this phase is due to exhaustion of nutrients, accumulation of toxic waste products, depletion of oxygen, or development of an unfavourable pH which are the factors responsible for the decline in the growth rate.

Death Phase

At some point, determined by a combination of factors, the stationary phase ends and net death begins. Death can be measured by determining the proportion of total cells that are alive at a particular time. The exact shape of the curve during the death phase will depend on the nature of the organism under observation and the many factors that contribute to cell death. The death phase may assume a linear function such as during heat-induced death, where viable cell numbers decline logarithmically.

Some additional considerations of the growth curve are important in assessing the effect of various internal as well as external factors on growth. The number of cells in a population (B) is equal to the number of cells in the initial inoculum (a) × 2^n,

$$b = a \times 2^n$$

Then

$$\log_2 b = \log_2 a + n$$
$$\log_{10} b = \log_{10} an + \log_{10} 12$$
$$\log_{10} b = \log_{10} a + (n \times 0.3010)$$

Solving the equation for n, the number of generations that occurred between the time of inoculation and the time of sampling is

$$n = \frac{\log_{10} b - \log_{10} a}{0.3010}$$

The generation time (tg) or doubling time may be determined by dividing the time elapsed (t) by the number of generations (n).

$$tg = t/n$$

MEASUREMENT OF CELL GROWTH

The growth of the cell can be measured by the following methods.

1. Total cell count and
2. Viable count

Total Cell Count

The samples are directly counted by placing under a counting chamber and viewed under the microscope. Then the number of cells are calculated and manipulated to the total volume of the sample. This type of count has certain limitations.

i. Dead cells are also counted.

ii. Some cells cannot be seen under the microscope.

iii. Not suitable for low-density samples less than 10^6/ml.

iv. Phase contrast microscope is needed.

Viable Count

This is used to count the colonies of the microorganisms. Each viable cell produces a colony. By this assumption two methods are generally applied. They are:

Spread plate technique   In this method the sample of 0.1 ml is poured on top of the agar and spread throughout the medium, using a glass spreader. After incubation the colonies are counted by direct counting or by using colony counter.

Pour plate technique   The sample of 0.1–1 ml is mixed with molten agar and poured in the petri plate. After incubation, colonies are counted. The temperature of the molten agar (45°C) which is added with it may sometimes cause the death of the organisms.

   In spread plate technique, some enormous count may occur due to the fused colonies and due to too large colonies and those organisms unable to produce a colony. To avoid this, a diluted sample of culture (0.1 ml) is inoculated in the petri plate and spread using a glass spreader. The number of colonies after incubation does not exceed 300. Several 10-fold dilutions of the sample are commonly used. For 10-fold dilution (10^{-1}) dilution, one can mix 0.5 ml of sample with 4.5 ml of diluent or 1 ml of sample with 9.0 ml of diluent. If a 100-fold dilution (10^{-2}) is needed,

0.05 ml can be mixed with 4.95 ml dilutent or 0.1 ml with 9.9 ml of diluent.

MEASURE OF MASS

It is frequently the case that in addition to measuring cells, we may measure protoplasm without regarding the cellular packages it is found in. During growth, the synthetic activities of the cell result in major increases in cell mass, if done with care. We may use mass increase as a measure of growth. The most frequently used measures of mass are dry weight, protein, and DNA.

Dry Weight

Determination of dry weight is not an easy one. However, under appropriate conditions, it is a useful measure of growth. The most common way in which dry weight is determined is to separate cells from their growth medium by centrifugation, wash the separated cells, dry the resultant at a temperature sufficient to dry the cells (normally about 70°C) and weigh the dry cells until a constant weight is obtained which gives the total dry weight. The temperature often used to remove water, 70°C, is sufficient to remove water without causing loss of organic material. It is possible to use a slightly higher temperature, but care must be taken to ensure that the temperature used to remove water is not high enough to damage cellular material. Treatment of the dry cells at substantially high temperatures (400 to 600°C) causes combustion of the organic cell compounds, leaving inorganic ash. Substraction of the ash weight from the total dry weight yields the total organic dry weight, which is a reasonable measure of growth. Dry weight measurements must be made with caution and are quite prone to errors. In particular, washing with fluids that are not isotonic with cells can lead to weight loss through lysis. In contrast, isotonic washing can lead to erroneously high weight as a function of the incomplete removal of washing solution. In addition to these concerns, non-specific precipitation of inorganic materials during centrifugation or filtration may occur. If this material is confused with protoplasm, erroneously high dry weight values may again be obtained. Finally, there are occasions when extensive accumulation of intracellular storage materials occurs, leading to dry weights that do not represent true protoplasm. All critical dry weight measurements should be accompanied by a study of the chemical nature of the dry weight so that its true significance can be assessed.

Protein

Protein is a reasonable measure of growth since it normally constitutes a majority (50–70%) of the organic cellular dry weight. The method by which we assess cell proteins depends upon our purpose and the amount of material involved. For microbial physiology, three methods of protein measurement are commonly used:

1. Biuret procedure
2. Folin reagent and
3. Coomassie blue-dye-binding assay

DNA

DNA is often used as an index of growth. Even when it is not used for this purpose, DNA is a frequently measured parameter. Typically, DNA is measured in one of the following three ways:

1. by the absorption of intact material at 260 nm,
2. by dexoyribose measurements, and
3. by phosphate measurements.

When 260 nm absorption is used to measure DNA, the DNA must be purified since all other components may absorb at 260 nm. The ratio of absorption at 280 and 260 nm is often used as an index of purity in preparations that contain protein. With highly purified DNA, 5 mg of DNA/ml gives an absorption of 0.10 at 260 nm with a 1-cm light path. If DNA is not measured by 260 nm absorption, it may be assessed by colorimetric deoxyribose or phosphate measurements and from the calculation of DNA content from the proportion that deoxyribose or phosphate constitutes of the total weights of a DNA nucleotide building block.

CONTINUOUS CULTURE

Bacteria are grown in "batch" culture in which a flask containing media is inoculated and growth is allowed to occur. This is a closed system in which the bacterium is very difficult to be manipulated for growth rate. In batch culture, growth rate is determined internally by properties of the bacteria themselves. During batch cultutre, the environment to which the cells are exposed is constantly changing. In early stages, nutrients are abundant and toxic products accumulation is minimal. As growth proceeds and cell numbers increase, the nutrient supply decreases and accumulation of end products becomes significant. Since the cell population is larger, the rate of nutrient depletion and toxic product formation both increase until the organism is, in effect, both starving and "drowning in its own garbage."

In continuous culture, the growth is unbalanced. This is often useful to study the growth of microbes in a way such that a constant environment is maintained. Continuous culture is achieved by simultaneously removing waste products and adding new nutrients to the system in a controlled manner. Continuous culture may be regarded as a means of maintaining microbes continually in the logarithm phase. By growing cells in such a fashion, three things are achieved:

1. A constant microbial population is present at all times.

2. The environment is constant.

3. The rate of growth is both constant and controlled.

In continuous culture systems, the volume in the culture vessel is kept constant by an overflow device that removes medium and cells at the same rate as fresh medium is *added.* In a chemostat, growth rate is determined externally by altering the rate-limiting nutrient added to the culture vessel. The faster the limiting nutrient is added, the faster is the growth rate.

Mechanisms of Continuous Culture Control

In continuous culture, regulation of the cell population, waste product removal and nutrient addition may be controlled in two ways, by "turbidostatic" control or by "chemostatic" control. In the former case, the turbidity, or cloudiness of the culture itself is used to control the system through the aid of a photocell. In chemostatic control of continuous culture, the population is controlled by a limiting nutrient which determines the growth rate of the culture, since all other required nutrients are supplied in excess.

Turbidostat   The first type of continuous culture system, the turbidostat, has a photocell that measures the absorbance or turbidity of the culture in the growth vessel. The flow rate of media through the vessel is automatically regulated to maintain a predetermined turbidity or cell density.

Chemostat   It is so constructed that the sterile medium is fed into the culture vessel at the same rate as the media containing microorganisms is removed. The culture medium for a chemostat possesses an essential nutrient (e.g. an amino acid) in limited quantities. Because of the presence of a limiting nutrient, the growth rate is determined by the rate at which the new medium is fed into the growth chamber, and the final cell density on the concentration of the limiting nutrient.

Merits

Continuous culture systems are very useful because

i.   they provide a constant supply of cells in the exponential phase and grow at a known rate.

ii. they make possible the study of microbial growth at very low nutrient levels, at concentrations close to that present in the natural environment.

SYNCHRONOUS CULTURE

Continuous culture allows maintenance of a constant population of cells at all times. Because of growth rate control and the balancing of growth rate with end product removal, the environment to which the cells are exposed is constant, as opposed to a batch culture, in which it is continually changing. In the continuous culture, the constancy of growth conditions is accompanied by constancy in cell composition because the coordinated processes of growth are affected in a constant way, and therefore, maintain a constant relationship to each other. On an average, the behaviour of a cell population in continuous culture is constant, although the behaviour of individual cells is not uniform. For certain purposes, particularly for the study of the relationship of events in the cell cycle to each other, it is useful if not essential, that all of the cells in a microbial population behave in the same way. Synchronous culture is the mechanism by which this objective is obtained. The graphical representation of synchronous outline is shown in Figure 6.2. The usefulness of such a procedure is exemplified by the knowledge that DNA replication is semiconservative. If it were not for the ability to achieve synchronous culture, or to closely approximate it, the demonstration of the semiconservative nature of DNA replication would not have been possible.

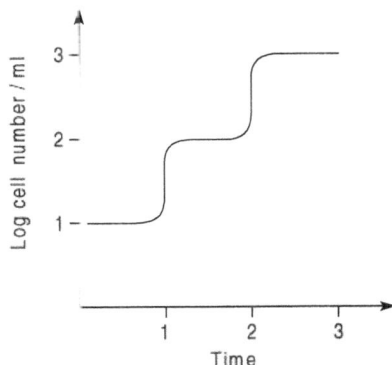

Figure 6.2   Graphical representation of synchronous culture (Growth).

An exponentially growing culture shows different stages of growth. To examine the physiology of each cell, (single) culture in that medium is a very difficult task. So a system that closely resembles and amplifies the behaviour of single cells in a synchronous culture which contains cells that are physiologically identical and are in the same stage of division or by forcing a cell population to attain an identical physiological condition by a change in the environment is effected. There are several methods to obtain synchronous culture by centrifugation, filtration or by periodic changes in nutritional and environmental conditions. In a synchronous culture since the cells are physiologically identical, cell division occurs periodically at constant intervals.

Other cell constituents such as dry mass, optical density, total protein or RNA content/cell increases at a constant rate but the amount/millilitre of culture will increase in population proportional to the cell number. On the other hand, the pattern of DNA synthesis can be either periodic or continuous depending on how fast the culture is growing. For example *E. coli* growing in a medium with a generation time of less than 40 min will show continuous DNA synthesis, while growing with the generation time greater than 40 min will show a period when no DNA synthesis occurs. Although such synchronous culture serves as an excellent amplification of a single cell system, their usefulness is restricted since perfect synchrony can be had only for 2–3 division cycles. In recent years, a further improvement has been made to obtain information regarding individual cell cycles. In this, a culture of cells growing exponentially is centrifuged in sucrose, glycerol or sorbitol gradients to separate cells based on their densities which is directly related to their age. Analysis of such fractions from gradient centrifugation has provided the same information as one would get from using synchronous culture.

A number of methods have been devised to approximate synchronous culture. These include: (i) nutrient limitation (ii) temperature oscillation and (iii) cell selection. In nutrient limitation, the strategy is to starve the cells to the point that growth is no longer possible and, at a particular time, to resupply the organism with the required nutrient.

Sometimes, achievement of synchrony in this manner is difficult because of the necessity of extreme depletion of intracellular reserves of the limiting nutrient.

STUDY OUTLINE

- Growth can be defined as the final expression of the physiology of an organism.

- For batch culture, there are four stages of growth.

- In the first stage, i.e. lag phase, otherwise called as adaptation phase, the organisms adapt themselves to the nutrients.

- In the second stage, i.e. log phase, the growth rate of the organisms is very high and they accumulate the primary metabolites in the medium.

- In the third stage, i.e. stationary phase, the growth is limited and is otherwise called as cryptic growth.

- In the last stage, i.e. death phase, because of the accumulation of secondary metabolites like alcohols, pigments, toxins, and antibiotics, the death of the organism occurs. In addition, nutrient depletion also leads to the death of the organism.

- Growth can be measured by various methods like, direct count, viable count, spread plate, and pour plate techniques.

- Some other types of growth that can be achieved in microorganisms include, continuous culture and synchronous culture methods.

- Continuous addition of nutrients and the same amount of removal of inoculums from the culture vessel leads to the even distribution of inoculum to the microorganisms.

CONCEPT CHECK

1. How can you measure the growth of microorganisms?

2. Write about the various stages of growth in batch culture.

3. Define growth. What is "sur" gene?

4. Give an outline about the advantages of continuous growth.

5. Elaborately write about the details of synchronous growth.

6. How will you calculate growth rate?

7. What is doubling time?

8. What are primary and secondary metabolites?

CRITICAL THINKING

In the natural environment, the growth of an organism takes place based on nutrient availability, but the organism does not know how much grams of amino acids, proteins or carbohydrates it needs? But in the laboratory, for the cultivation of microorganisms, we are supplying a media at a constant rate. On what basis can one formulate the requirements of a microorganism?

7

INFLUENCE OF ENVIRONMENTAL FACTORS ON GROWTH

INTRODUCTION

The growth of the organisms is greatly affected by the physical and chemical nature of their surroundings. Most organisms can survive in moderate environments. Some organisms can inhabit extreme environments. For example *Bacillus infernus* can live at 1.5 miles below the earth in anaerobic conditions at 60°C. This type of organisms are called as extremophiles. The following are the factors which mainly influence growth.

pH

pH is a measure of the hydrogen ion concentration of a solute and is defined as the negative logarithm of the hydrogen ion concentration, expressed in terms of molarity.

$$pH = -\log[H^+] = \log[1/H^+]$$

The pH scale extends from pH 0.0 to pH 14.0 and each pH unit represents a 10-fold change in hydrogen ion concentration. pH dramatically affects microbial growth. Depending on the pH, microorganisms are of three types.

Acidophiles   The name implies those organisms that have their growth between pH 0 and pH 5.5. e.g. *Sulfolobus acidocaldarius* are common inhabitants of acidic hot springs, grow well at pH 1-3 and at high temperatures.

Neutrophiles   These organisms have their growth between pH 5.5 and 8.0. *E. coli* has this range of growth and tolerability. The optimum pH for *E. coli* is pH 7.0.

Alkalophiles   Organisms of this type have their growth between pH 8.5 to 11.5. They are so called because they can tolerate the alkaline environment, e.g. *Vibrio cholerae*, whose optimum pH is 10 or higher.

The drastic change in the pH range can harm microorganisms by disturbing the plasma membrane or inhibiting the activity of enzymes and membrane transport protein. Bacterial death occurs if the internal pH drops much below 5.0–5.5. So the change in pH might also alter the ionization of the nutrient molecule and thus reduce the availability to the organism.

Despite wide range variations in the pH of the habitat, the internal pH of most microorganisms is close to neutrality. The microorganisms have some adaptation mechanisms for maintaining the pH at neutrality, and these are listed below.

i.   The plasma membrane may be relatively impermeable to protein.

ii. Neutrophiles appear to exchange potassium ions, for proteins are antiport systems (a type of transport system in which the transported substances move in opposite direction). e.g. in *E. coli*, sodium transport system pumps sodium outward in response to the inward movement of proteins.

iii. If the pH becomes too acidic, other mechanisms come into play. When the pH drops below 5.5–6.0, the organism synthesizes new protein for its survival. e.g. *Salmonella typhimurium* and *E. coli* synthesize an array of new proteins as part of which has been called their acidic tolerance response.

iv. If the external pH decreases to 4.5 or lower, chaperones such as heat or acid-shock proteins are synthesized.

v.   Microorganisms frequently change pH of their own habitat by producing acidic or basic metabolic waste products.

vi. In the case of in vitro conditions, buffers are often included in the media to prevent growth inhibition by large pH changes. Phosphate is a commonly used buffer and a good example of buffering by a weak acid ($H_2PO_4^-$) and its conjugative base (HPO_4^{2-}).

TEMPERATURE

Microorganims are unicellular in nature. So they show high degree of diversity in their temperature range. Because of their membraneless components, they are highly susceptible to temperature. The term poikilothermic denotes that their temperature varies with that of external environment. Each organism has minimum, optimum and maximum temperature range termed as cardinal temperature.

The cardinal temperature may vary greatly between microorganisms. However their optima normally ranges from 0°C to as high as 75°C, whereas microbial growth occurs at temperatures extending from –20°C to over 100°C. The microorganisms which are having a small range of growth temperature are termed as stenothermal, e.g. *Neisseria gonorrhoea*. Eurythermal microorganisms have a wide range of growth temperature and can survive various ranges of temperature, e.g. *Enterococcus faecalis*. But in the case of protrozoans, their growth temperature is 50°C while some algae and fungi can grow at 55–60°C. But bacteria can grow at 100°C. The reason for susceptibility of eukaryotes and prokaryotes is their inability to synthesize their cell membranes at 60°C.

Based on their optimum temperature range, microorganisms can be classified into five classes:

1. *Psychropiles* are those organisms which can grow well at 0°C and have an optimum growth at 15°C or lower. The maximum temperature is around 20°C. The organism can adapt to this low temperature because the cell membrane has high levels of unsaturated fatty acids and remains semifluid when cold. The term 'psychrotrophs' or facultative 'psychrotroph' refers to organisms which can grow even at 0°C, even though they have an optima between 20 and 30°C, and maxima of about 37°C. Psychrotrophic bacteria are the major causatives of spoilage of refrigerated food.

2. *Mesophiles* are those microorganisms with growth optima around 20–45° and a temperature minimum of 15–20°C. Their maximum is about 45°C. Almost all human pathogens are mesophiles. e.g. *E. coli, Salmonella*, etc.

3. *Thermophiles* are organisms which can grow at temperatures of 55°C or higher. Their growth minimum is usually around 45°C and they often have optimum between 55°C and 65°C.

These organisms adapt themselves at high temperatures due to the presence of stable enzymes and protein synthesis systems able to function at high temperature. Their membrane lipids are more saturated and have higher melting points.

4. *Hyperthermophiles* are those organisms that can grow at 90°C or above and some have maxima above 100°C. Bacteria have growth optima between 80°C and about 113°C are called as hyperthermophiles, e.g. *Bacillus staerothermophilus*.

RADIATION

Radiation plays a major role in growth. Radiations of longer wavelength have lesser energy whereas those with shorter wavelength have greater energy. Electromagnetic radiations also act like a stream of energy packets called photons. Each photon has a quantum of energy whose value will depend on the wavelength of the radiation. Sunlight is the major source of radiation on the earth. It includes visible light, ultraviolet rays, infrared rays, and radio waves.

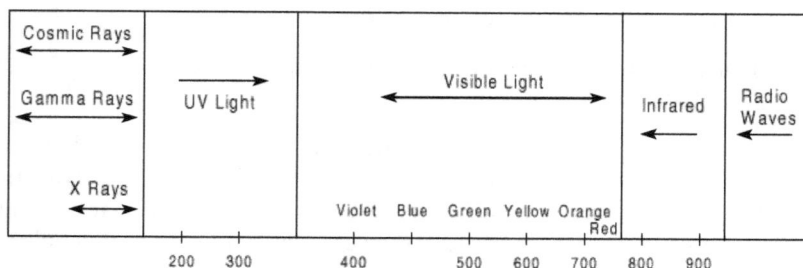

Cosmic Rays ◄──►					
Gamma Rays ◄────►	UV Light ────►	Visible Light ◄──────────────►		Infrared ◄──	Radio Waves ◄──
X Rays ◄──►		Violet　Blue　Green　Yellow　Orange　Red			

```
        200   300      400      500  600  700  800  900
```

Ionizing Radiation

These are radiations of shorter wavelength or high energy which can cause atoms to lose electrons or ionize. Two major ionizing radiations are (i) X-rays, which can be artificially produced and (ii) Gamma rays, which are emitted during decay of radioisotopes. These rays cause mutation. But some microorganisms can tolerate and grow in the presence of radiation, e.g. *Deinococcus radiodurans*. This radiation breaks hydrogen bonds, oxidizes double bonds, destroys ring structures, polymerizes some molecules and ($OH^\bullet$) radical formation. Ionizing radiation also causes the destruction of DNA and leads to the death of the microorganism.

Ultraviolet Radiation (10 to 400 nm)

Ultraviolet radiation is most lethal at 260 nm. Because DNA absorbs the UV light at this wavelength only, they cause thymine dimer in adjacent base of the DNA strand. This damage can be repaired by photoreactivation with the help of the enzyme photolyase.

Visible Light

Visible light is beneficial in photosynthesis. Yet even visible light when present in sufficient level can cause damage to the cell. When visible light is absorbed by the photosensitizers (like bacterial chlorophyll, cytochrome, and flavanoids), they get excited and transfer O_2 (singlet oxygen $1O_2$) to the cell and immediately cause damage of the cell.

STUDY OUTLINE

- Growth of the microorganisms is influenced by various factors like temperature, pH, radiation, etc.

- Some microorganisms can tolerate extreme environmental conditions so they are termed as extremophiles.

- Based on the pH, the organisms are classified into acidophiles, neutrophiles, and alkalophiles.

- Each organism has an adaptation mechanism to overcome this factor.

- Based on the temperature, they are classified into psychrophiles, mesophiles, thermophiles, and hyperthermophiles.

- The term poikilothermic indicates that the internal temperature of microorganisms varies with that of the external environment.

- Microorganisms have different temperature range termed as cardinal temperature which denotes, minimum, optimum, and maximum growth temperature.

- The microorganisms which have a small range of growth temperature are termed as stenothermal.

- Radiation also plays an important role in the growth of microorganism.

- In ionizing radiation, radiation of shorter wavelength or higher energy can cause atoms to lose electrons or ionize.

- Even visible light which is beneficial in photosynthesis, when present in sufficient quantities, can damage the cell.

CONCEPT CHECK

1. Define acidophiles, neutrophiles, and alkalophiles.
2. What is poikilothermic?
3. Define stenothermal and cardinal temperature?
4. What are photons?
5. How do organisms maintain themselves in extreme environnments?

CRITICAL THINKING

Microorganisms adapt themselves for the extreme environment where they grow. Is there any possibility of adaptation of a microorganism to switch over from one extreme environment to another, as for example from acidophilic environment to alkalophilic environment?

8

BACTERIAL ENZYMES

INTRODUCTION

"Enzymes" are proteins which are biological catalysts which have great specificity for the reaction catalysed and the molecules acted on. An enzyme is a substance that increases the rate of a reaction without itself being altered or consumed in the reaction. In other words, an enzyme decreases the activation energy needed to initiate the reaction.

In microorganisms, the enzymes are synthesized during biochemical reactions. For each substrate, an activating enzyme should be there in the bacterium. For example, *E. coli* contains over 1000 different enzymes that decrease all its physical characteristics and dictate its metabolic activity. Some organisms produce enzymes which are responsible for the pathogenicity of the particular organism. For example, the bacterium *Clostridium perfringens* has an enzyme called lecithinase which cleaves the lecithin in the cell membrane of the host tissue and causes damage to the host cell, which leads to the deadly disease "gas gangrene." Some organisms produce some enzymes which alter the metabolic pathway of the host. For example *Cornyebacterium diptheriae* produces toxin-like enzymes which inhibits protein synthesis of the host. So it causes the death of the host due to lack of proteins.

Enzymes speed up the chemical reactions. The reactants are called as substrates and the substance formed from the reaction is the product.

PROPERTIES OF ENZYMES

The characteristic features of enzymes may be listed as follows:

    i.   Simple enzymes are made up of a protein part alone and they are otherwise called as "apoenzymes." Complex enzymes, are made up of a protein part and a small molecule of nonprotein. They are otherwise termed as "holoenzyme" (collectively).

ii.   If a protein and nonprotein part (cofactor) bind tightly and mediate a reaction, it is called as "prosthetic group."

iii. Enzymes are highly specific in their catalytic action, i.e., an enzyme recognizes only one set of substrate (the material on which it will react) and converts it to one particular set of product.

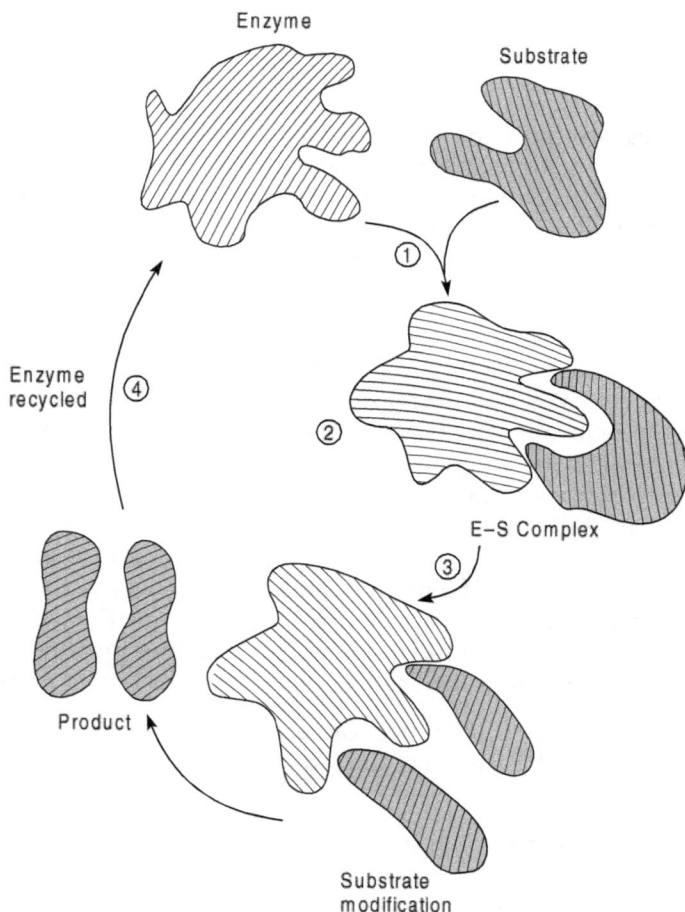

Figure 8.1   Mechanism of action of simple enzyme.

iv. The specificity of the enzyme induces complementary three-dimensional shapes to the two molecules.

v. The enzyme contains an area called the active site that reacts with the substrate. The substrate must fit precisely into the active site, in much the same way as a key fits into a lock.

vi. If the shapes are not complementary, nothing happens. The high degree of specificity enables the cell to control the metabolic reactions by regulating the type of enzyme produced.

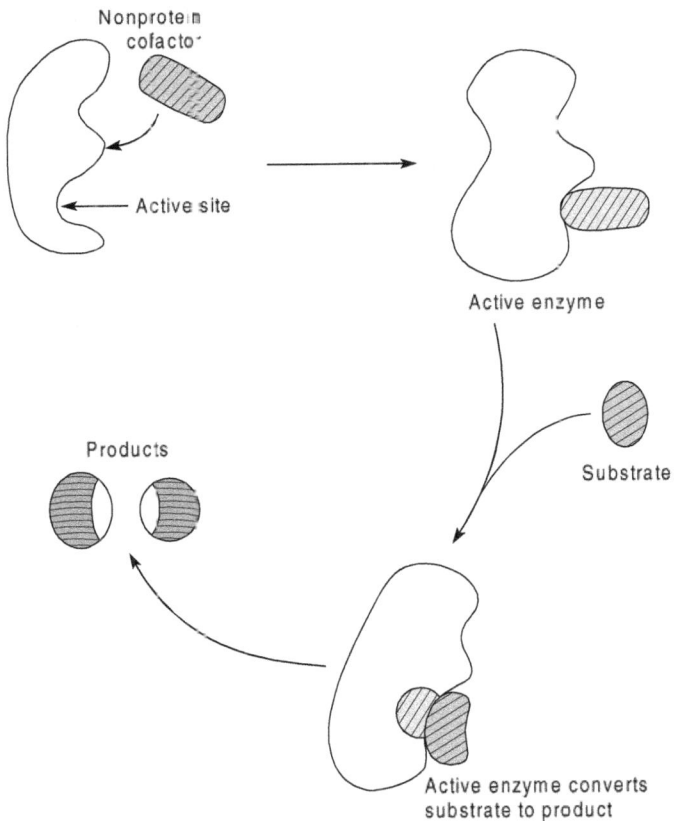

Figure 8.2 Mechanism of action of complex enzyme.

vii. Enzymes are very active even at low concentrations. Even very low concentration of enzyme can induce reactions rapidly.

viii. Enzymes are never consumed in the reaction. They recycle after the product is formed. This recycling of enzymes contributes to their efficiency.

During the simple-enzyme-mediated chemical reaction, (Figure 8.1) the enzyme first binds with substrate which forms enzyme–substrate complex. This is an intermediate state where the substrate is going to convert as product. After the conformational change of the substrate due to the binding of enzyme, it modifies the substrate, as a result the product is formed. In the case of complex-enzyme-mediated reaction (Figure 8.2), a, non-protein part is needed. This non-protein part speeds up the rate of reaction and mainly acts as the electron carrier of the reaction. When the enzyme binds with non-protein part it become a complete enzyme. The simple-enzyme part i.e., protein part of an enzyme has the active site, whereas the non-protein part binds. This stimulates the enzyme to be active. When this active enzyme binds with substrate, it converts the substrate into product.

CLASSIFICATION OF ENZYMES

In 1961, the International Union of Biochemists (IUB) adopted a new system of classification and nomenclature of the enzymes. According to this system, enzymes have been classified into six major classes.

1.   Oxidoreductases
2.   Transferases
3.   Hydrolases
4.   Lyases
5.   Isomerases
6.   Ligases

Each enzyme has the ability to mediate or catalyse a type of reaction.

Oxidoreductases

These enzymes have the potential to bring the oxidation–reduction reactions. Such enzymes can oxidize their substrate by addition of oxygen, or by removal of hydrogen or electrons. Reduction is effected by the opposite reactions that occur during oxidation, e.g. cytochrome oxidase, glutamate dehydrogenase, catalase, peroxidase, etc. The following reaction is an example for oxidation–reduction of an enzyme.

Lactate dehydrogenase

$$\text{Pyruvate} + \text{NADH} + \text{H}^+ \rightleftharpoons \text{lactate} + \text{NAD}^+$$

Transferases

Enzymes which bring about transfer of a group other than hydrogen from one substrate to another are known as transferases. These groups transfer the functional molecules which include, amino, methyl, formyl, acyl, glycosyl, phosphate, thiol, ketol, aldol, etc. Examples for this group of enzymes are transaminase, transmethylase, transformylase, transketolase, etc.

Aspartate carbamoyl transferase

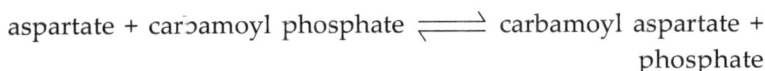

$$\text{aspartate} + \text{carbamoyl phosphate} \rightleftharpoons \text{carbamoyl aspartate} + \text{phosphate}$$

Hydrolases

Enzymes which catalyse the hydrolysis of the substrate by adding constituents of water across the bond they split are known as hydrolases. The substrates might be esters, peptides, glycosides, etc. e.g. lipases, phosphatases, amidases, deaminases, nucleases, pepsin, trypsin, chymotrypsin, cholinesterase, urease, glutaminase, etc.

Glucose 6-phosphatase

$$\text{glucose 6-phosphate} + \text{H}_2\text{O} \longrightarrow \text{glucose} + \text{phosphate}$$

Lyases (Demolases)

These enzymes catalyse the removal or addition of groups to the substrate which have double bond. The enzyme also mediates the lysis of substrate other then hydrolysis. Examples for this group of enzymes are fumarase, aldolase, various decarboxylases, carbonic anhydrase, aspartase, lyase, etc.

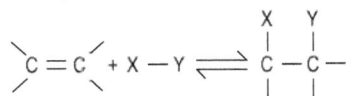

$$\overset{\diagdown}{\underset{\diagup}{\text{C}}} = \overset{\diagup}{\underset{\diagdown}{\text{C}}} + \text{X} - \text{Y} \rightleftharpoons \overset{\text{X}}{\underset{|}{\text{C}}} - \overset{\text{Y}}{\underset{|}{\text{C}}} -$$

Fumarase hydratase

$$\text{L-malate} \rightleftharpoons \text{Fumarate} + \text{H}_2\text{O}$$

Isomerase

This group of enzymes catalyse those reactions in which the substrate is converted into the positional or optical isomer by intramolecular arrangements of various atoms or groups. Examples for this group of enzymes are aconitase, phosphohexose isomerase, triose phosphate isomerase, phosphohexomutase, racemase, etc.

$$\text{L-Alanine} \rightleftharpoons \text{D-Alanine}$$

Ligases (Synthetase)

These enzymes catalyse the linking together of compounds utilizing energy made available by simultaneous breakdown of ATP or similar high-energy compounds. Examples for this type of enzymes are succinic thiokinase, glutamate synthetase, acetyl CoA carboxylase, etc. Joining of two molecules using ATP energy (or that of other nucleoside triphosphate).

Glutamine synthetase

$$\text{glutamate} + NH_3 + ATP \longrightarrow \text{glutamine} + ADP + Pi$$

ENZYME NOMENCLATURE

The international system of naming enzymes by adding the suffix 'ase' to the end of either the name of the substrate or the name of the reaction catalysed, is called as enzyme nomenclature. The name of the enzyme generally consists of two parts. The first part denotes the name of the substrate and the second part denotes the type of reaction catalysed. Each enzyme is given a systematic code number which denotes the class, subclass and the name of the enzyme. For example, in the number 2.7. 1.1, 2 denotes the class transferase, 7 is the subclass which denotes the transfer of phosphate, 1 denotes the alcohol working as phosphate acceptor and the last 1 denotes the name of the enzyme hexokinase.

Glutamate dehydrogenase	1.4.1.3 L-glutamate NADP oxidoreductase
Fumarase	4.2.1.2 L-malate hydrolyase
Succinate thiokinase	6.2.1.4 Succinate CoA ligase

ISOENZYMES

Recent studies have shown that some of the enzymes exist in many physically distinct forms and all the forms can catalyse the same reaction. The only differences noted in the different forms of a single enzyme is that of electrophoretic mobility and possibility of the quaternary structure of the enzyme protein. Such forms of a single enzyme, having the same catalytic activity are known as multiple forms of an enzyme occurring within a single species. Some 100 enzymes at present are known to exixt in such multiple forms, e.g. malate dehydrogenase.

ADAPTIVE AND CONSTITUTIVE ENZYMES

These are two major classes of enzymes in which the adaptive enzyme formation is dependent upon the adaptation of the organism to some nonphysiological substances. Such enzymes are produced only if suitable substrates are present, e.g. β-galactosidase.

In the case of constitutive enzymes, the synthesis is independent. They are synthesized during metabolism both in the absence or presence of a substrate, e.g. enzymes in glycolysis.

Table 8.1 Coenzyme activity of B-Complex.

Vitamin	Coenzyme	Type of reaction assisted by coenzyme
Thiamine	TPP	Removal of CO_2 from molecules (decarboxylation)
Riboflavin (B_2)	FDD	Hydrogen carriers in energy-generating reactions
Pyridine (B_6)	Pyridoxyl phosphate	NH_2 transfer (transamination), decarboxylation of amino acids, removal of H_2O from amino acids
Cobalamine (B_{12})	Cobalamine	Protein and nucleic acid metabolism

Table 8.1   Contd.

Vitamin	Coenzyme	Type of reaction assisted by coenzyme
Niacin (nicotinic acid)	NAD, NADP	Carriers of H^+ in energy-generating reactions and biosynthesis
Pantothenic acid (B_5)	Coenzyme A	Transfer of small organic molecular fragments in respiration and fatty acid metabolism
Folicin	Folic acid	Single carbon transfer in nucleic acid and amino acid metabolism.

STUDY OUTLINE

- Enzymes are protein biological catalysts which have great specificity for the reaction catalysed and the molecules acted upon.

- In microorganisms, the enzymes are synthesized during biochemical reactions.

- Based on the properties, the enzymes can be classified into two types (i) simple enzymes (ii) complex enzymes.

- Simple enzyme is otherwise called as apoenzyme and complex enzyme as holoenzyme.

- Enzymes are classified into 6 classes. The classification is based on the activation of the enzyme and the mechanism of enzyme action.

CONCEPT CHECK

1. Define simple and complex enzymes.
2. Write about the properties of enzymes.
3. How will you name an enzyme?
4. What are constitutive enzymes and adaptive enzymes.
5. Define an isoenzyme with an example.
6. Explain the six major classes of enzymes with examples.

CIRTICAL THINKING

For each substrate, specific enzymes are provided in a system. How does the system know that the particular substrate is entering?

9

GLYCOLYSIS OR EMBDEN–MEYER PATHWAY

Glycolysis is an almost universal central pathway of glucose catabolism. It is the pathway through which the longest flux of carbon occurs in most cells. It is a major pathway for the conversion of hexose sugars to pyruvate. A molecule of glucose is degraded in a series of enzyme-catalysed reactions to yield 2 molecules of pyruvate. During the sequential reaction of glycolysis, some of the free energy released from glucose is conserved in the form of ATP. The operation of the pathway results in the formation of two molecules of reduced nicotinamide adenine dinucleotide (NADH) and two molecules of ATP per molecule of glucose degraded to pyruvate. The pathway results in the formation of six of the critical biosynthetic intermediates, glucose 6-phosphate, fructose 6-phosphate, triose phosphate, 3-phosphoglycerate, phosphoenol pyruvate (PEP), and pyruvate.

Many type of anaerobic microorganisms are entirely dependent on glycolysis. The pathway is symmetrical since cleavage of the glucose molecule between carbons 3 and 4 converts a six-carbon molecule into two three-carbon molecules in such a way that each carbon at the top half of the glucose molecule has a metabolically equivalent carbon at the bottom half of the molecule. Thus, the phosphorylated carbons in triose molecules arise from glucose carbons 1 and 6, the carbons adjacent to the phosphate of the trioses arise from glucose carbons 2 and 5, and the carbons distal from the phosphated carbons of trioses arise from glucose carbons 3 and 4. Although the pathway generates NADH, it does not generate NADPH. So it cannot, by itself, provide reducing power for biosynthesis.

AN OVERVIEW OF GLYCOLYSIS

Glycolysis consists of two stages:

1. Preparatory phase (or) energy investment phase and
2. Redoxy reaction (or) energy generating phase

Preparatory Phase

This phase is called energy investment phase because two ATP molecules are consumed in this phase. The need of energy for this reaction is satisfied by the ATP which is then released as ADP.

The reaction involves various steps as follows:

Reaction I   The first ATP investment

Reaction 1 begins with ATP-dependent phosphorylation of glucose, catalysed by hexokinase. Magnesium ion is required, because the reactive form of ATP is its chelated complex with Mg^{2+}. This is true for virtually all ATP-requiring enzymes.

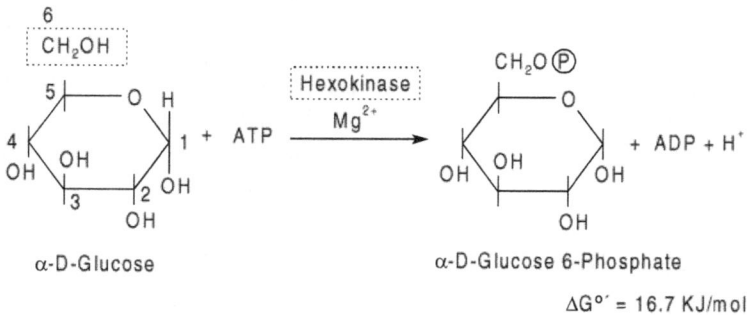

α-D-Glucose   α-D-Glucose 6-Phosphate

$\Delta G^{\circ\prime} = 16.7$ KJ/mol

Reaction II   Isomerization of glucose 6-phosphate

The net reaction, catalysed by phosphoglucoisomerase, is the readily reversible isomerization of the aldose sugar glucose 6-phosphate (G_6P) to the corresponding ketose sugar fructose 6-phosphate (F_6P). The effect of transferring the carbonyl oxygen from carbon 1 to carbon 2 is that the hydroxyl group generated at carbon 1 can be readily phosphorylated in the next reaction.

CH$_2$O (P)

Phosphoglucoisomerase

α-D-Glucose 6-phosphate

(P)OCH$_2$ CH$_2$OH

D-Fructose 6-phosphate

$\Delta G^{\circ\prime} = + 1.7$ KJ/mol

CH$_2$O (P)

Glucose 6-phosphate ⇌   ⇌ Fructose 6-phosphate

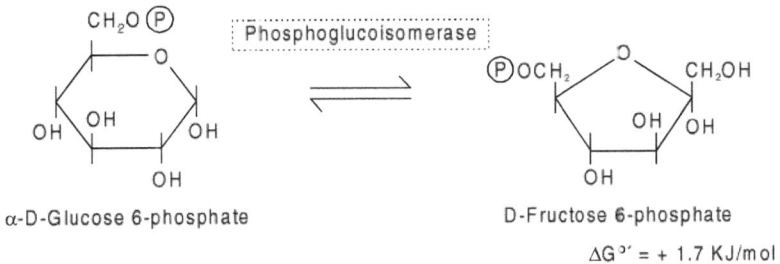

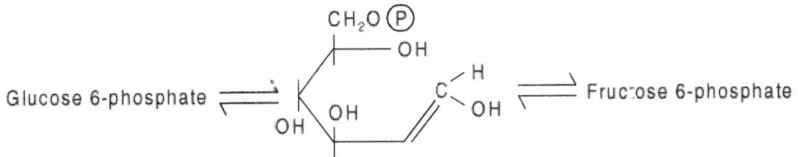

Reaction III   The second ATP investment

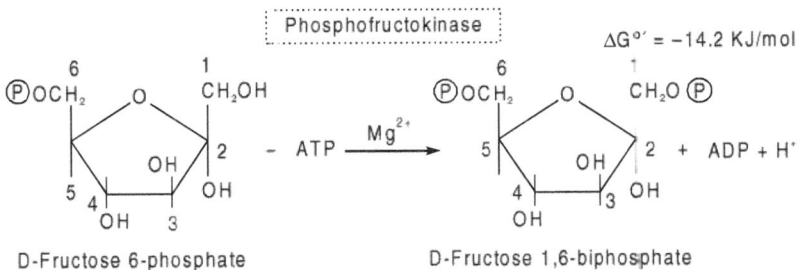

Phosphofructokinase

$\Delta G^{\circ\prime} = -14.2$ KJ/mol

(P)OCH$_2$ CH$_2$OH $- \text{ATP} \xrightarrow{\text{Mg}^{2+}}$ (P)OCH$_2$ CH$_2$O (P) $+ \text{ADP} + \text{H}^{\cdot}$

D-Fructose 6-phosphate   D-Fructose 1,6-biphosphate

The enzyme phosphofructokinase mediates the ATP-dependent phosphorylation. The product is called as fructose 1,6-biphosphate or fructose 1,6-diphosphate. The renaming was done to show that the two phosphates are separate, rather linked in ADP.

The enzyme phosphofructokinase is an allosteric enzyme whose activity is sensitive to the energy status of the cell, as well as to the levels of various other intermediates.

Reaction IV   Cleavage of two triose phosphate

The enzyme fructose 1,6-biphosphate aldolase, usually called aldolase is so named because its reaction is similar to the reverse of an aldol condensation. In this reaction, the splitting of sugars that is denoted by the term glycolysis occurs, because the six-carbon compound fructose 1, 6-biphosphate is cleaved to give two three-carbon intermediates. This reaction is strongly "endergonic". Aldolase

$$\text{(P)}OH_2C \diagdown O \diagdown CH_2O\,\text{(P)}$$

*D-Fructose 1,
6-biphosphate*

Fructose
1,6-biphosphate
aldolase

$$\begin{array}{l} CH_2O\,\text{(P)} \\ | \\ C=O \\ | \\ CH_2OH \end{array} \quad + \quad \begin{array}{l} O \\ \| \\ C-H \\ | \\ H-C-OH \\ | \\ CH_2O\,\text{(P)} \end{array}$$

*Dihydroxy
acetone phosphate*   *D-glyceraldehyde
3-phosphate*

$\Delta G^{\circ\prime} = +23.9 \text{ KJ/mol}$

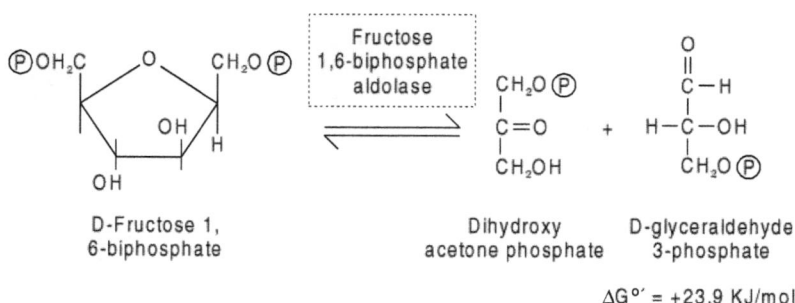

activates the substrates for cleavage by condensing the keto carbon at position 2 with lysine Σ-amino group in the active site to give a *Schiff Phase*. Schiff phase is a condensation product between an amino group and a carbonyl group. The activated substrate undergoes abstraction of a proton from the hydroxyl group at carbon 4, followed by an elimination of the resulting enolate ion, which results in splitting the bond between C-3 and C-4.

Reaction V   Isomerization of dihydroxyacetone phosphate

The enzyme triose phosphate isomerase converts the dihydroxy acetone phosphate to glyceraldehyde 3-phosphate, the substrate for the next glycolytic reaction. This reaction is also an endergonic reaction.

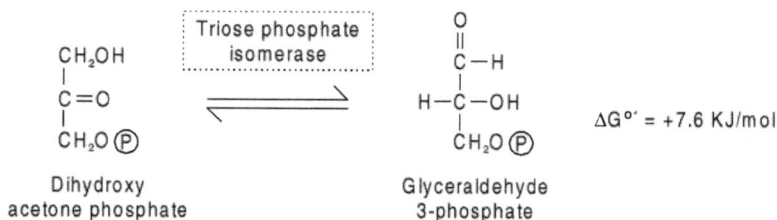

$$\begin{array}{l} CH_2OH \\ | \\ C=O \\ | \\ CH_2O\,\text{(P)} \end{array}$$

Triose phosphate
isomerase

$$\begin{array}{l} O \\ \| \\ C-H \\ | \\ H-C-OH \\ | \\ CH_2O\,\text{(P)} \end{array} \qquad \Delta G^{\circ\prime} = +7.6 \text{ KJ/mol}$$

*Dihydroxy
acetone phosphate*

*Glyceraldehyde
3-phosphate*

Reactions VI to X form the energy generation phase.

Reaction VI   Formation of NADH$_2$

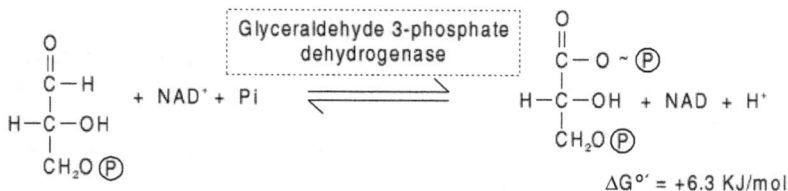

$$\begin{array}{l} O \\ \| \\ C-H \\ | \\ H-C-OH \\ | \\ CH_2O\,\text{(P)} \end{array} \quad + \text{ NAD}^+ + \text{ Pi}$$

Glyceraldehyde 3-phosphate
dehydrogenase

$$\begin{array}{l} O \\ \| \\ C-O \sim \text{(P)} \\ | \\ H-C-OH \\ | \\ CH_2O\,\text{(P)} \end{array} + \text{ NAD} + \text{ H}^+$$

$\Delta G^{\circ\prime} = +6.3 \text{ KJ/mol}$

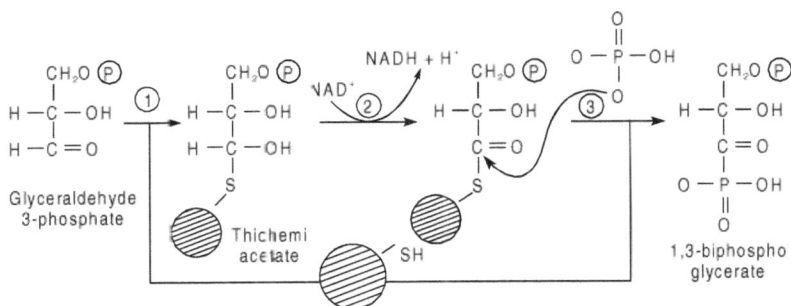

This reaction is the most interesting reaction among the glycolytic reactions which is catalysed by glyceraldehyde 3-phosphate dehydrogenase because it partly generates the first high-energy intermediate and partly a pair of reducing equivalents. The reaction involves a two-electron oxidation of the carbonyl carbon of glyceraldehyde 3-phosphate to the carboxyl level and the reaction is highly exergonic but slightly endergonic under standard conditions. This 1,3-bisphosphate has a phosphate on the carboxylic phosphoric acid anhydride, or an acyl phosphate group at position 1. The acyl phosphate is a high-energy phosphate group than the carboxylic anhydride group. This enzyme also requires a coenzyme, NAD^+, to accept electrons from the substrate being oxidized.

Reaction VII   The first substrate level phosphorylation

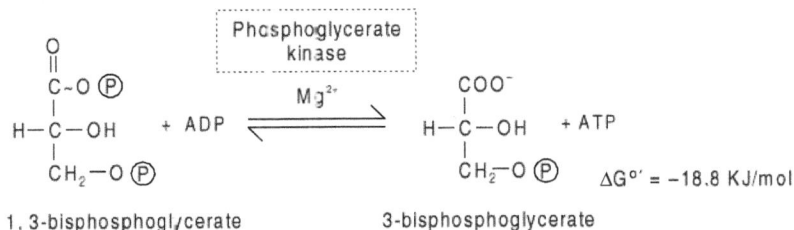

The acyl phosphate group in the 1,3 biphosphogycerate, because of its high group transfer potential, has a strong tendency to transfer its acylphosphate group with the resultant formation of ATP. At this stage, the ATP yield in the glycolysis pathway is zero. Recall that 2 moles of glucose are utilized to form 2 moles of triose phosphate. But in this reaction one molecule of ATP is formed.

Reaction VIII   Preparing for synthesis of the next high-energy compound

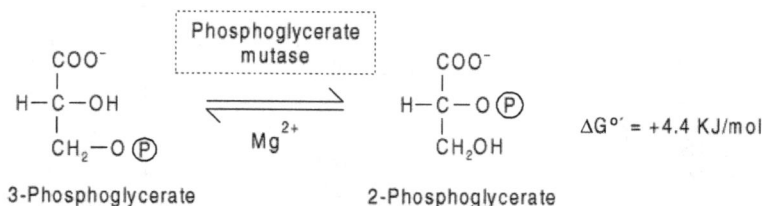

$$
\begin{array}{ccc}
\text{COO}^- & \boxed{\text{Phosphoglycerate mutase}} & \text{COO}^- \\
| & & | \\
\text{H}-\text{C}-\text{OH} & \xrightleftharpoons{\hspace{1.2cm}} & \text{H}-\text{C}-\text{O}\,\textcircled{P} \\
| & \text{Mg}^{2+} & | \quad \Delta G^{\circ\prime} = +4.4 \text{ KJ/mol} \\
\text{CH}_2-\text{O}\,\textcircled{P} & & \text{CH}_2\text{OH}
\end{array}
$$

3-Phosphoglycerate   2-Phosphoglycerate

The enzyme phosphoglycerate mutase causes the isomerization of the substrate and forms the product, 2-phosphoglycerate. The reaction is slightly endergonic. The enzyme contains phosphohistidine in its active site. In the first step of the reaction, phosphate is transferred to the substrate to give an intermediate 2,3-bisphosphoglycerate.

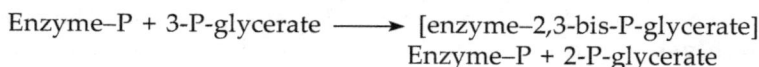

Enzyme–P + 3-P-glycerate $\longrightarrow$ [enzyme–2,3-bis-P-glycerate]

Enzyme–P + 2-P-glycerate

Reaction IX   Synthesis of second high-energy compound

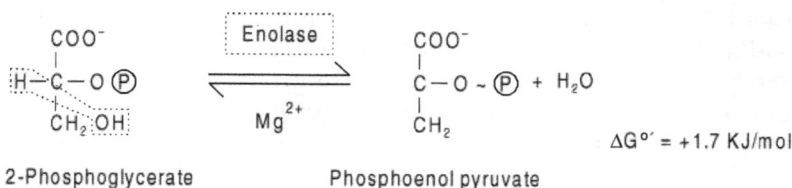

$$
\begin{array}{ccc}
\text{COO}^- & \boxed{\text{Enolase}} & \text{COO}^- \\
| & & | \\
\text{H}-\text{C}-\text{O}\,\textcircled{P} & \xrightleftharpoons{\hspace{1.2cm}} & \text{C}-\text{O} \sim \textcircled{P} + \text{H}_2\text{O} \\
| & \text{Mg}^{2+} & | \\
\text{CH}_2\text{OH} & & \text{CH}_2 \\
& & \quad \Delta G^{\circ\prime} = +1.7 \text{ KJ/mol}
\end{array}
$$

2-Phosphoglycerate   Phosphoenol pyruvate

The enzyme enolase generates another high-energy compound phosphoenol pyruvate (PEP) which participates in the second substrate-level phosphorylation of glucose. The reaction involves a simple dehydration, or α, β elimination and the overall free energy change is small.

Table 9.1   The critical biosynthetic metabolites.

Name	Structure	Major biosynthetic function
1. Glucose 6-phosphate	$\begin{array}{l}\text{CHO} \\ \mid \\ \text{H}-\text{C}-\text{OH} \\ \mid \\ \text{HO}-\text{C}-\text{H} \\ \mid \\ \text{H}-\text{C}-\text{OH} \\ \mid \\ \text{H}-\text{C}-\text{OH} \\ \mid \\ \text{CH}_2\text{O}-\text{PO}_3\text{H}_2 \end{array}$	Central intermediate in many catabolic pathways; carbohydrate polymers

Table 9.1 Contd.

	Name	Structure	Major biosynthetic function
2.	Fructose 6-phosphate	CH_2OH $\|$ $C=O$ $\|$ $HO-C-H$ $\|$ $H-C-OH$ $\|$ $H-C-OH$ $\|$ $CH_2O-PO_3H_2$	*N*-acetylglucosamine, *N*-acetylmuramic acid, murein
3.	Ribose 5-phosphate	$H\diagdown\diagup O$ C $\|$ $H-C-OH$ $\|$ $H-C-OH$ $\|$ $H-C-OH$ $\|$ $CH_2OPO_3H_2$	Nucleic acids, histidine
4.	Erythrose 4-phosphate	$H\diagdown\diagup O$ C $\|$ $H-C-OH$ $\|$ $H-C-OH$ $\|$ $CH_2OPO_3H_2$	Aromatic amino acids
5.	Triose phosphate	$H\diagdown\diagup O$ C $\|$ $H-C-OH$ $\|$ $CH_2O-PO_3H_2$	Precursor to dihydroxyacetone; lipids
6	3-phospho-glycerate	$\diagup O$ C $\|\diagdown OH$ $H-C-OH$ $\|$ $CH_2O-PO_3H_2$	Beginning point for the serine family of amino acids
7.	Phosphoenol pyruvic acid	$\diagup O$ C $\|\diagdown OH$ $C-O-PO_3H_2$ $\|$ C $H\diagup\diagdown H$	Key intermediate in several carbohydrate degradation sequences; aromatic amino acid precursor

Table 9.1   Contd.

Name	Structure	Major biosynthetic function
8. Pyruvic acid	$CH_3-C(=O)-C(=O)OH$	Key intermediate in degradative metabolism; "gateway compound" to the TCA cycle; precursor to the pyruvate family of amino acids
9. Acetyl-CoA	$CH_3-C(=O)-S-CoA$	Key energy compound; beginning point for lipid synthesis; participant in many synthetic processes
10. Alpha-ketoglutaric acid	$C(=O)OH$ — $C=O$ — $H-C-H$ — $H-C-H$ — $C(=O)OH$	TCA cycle intermediate; precursor to the glutamate family of amino acids
11. Oxaloacetic acid	$HO-C(=O)-CH_2-C(=O)-C(=O)OH$	TCA cycle intermediate; precursor to the aspartic family of amino acids
12. Succinyl-CoA precursor to heme	$C(=O)-S-CoA$ — CH_2 — CH_2 — $COOH$	Key intermediate in the TCA cycle; major energy intermediate

Reaction X The second substrate level phosphorylation

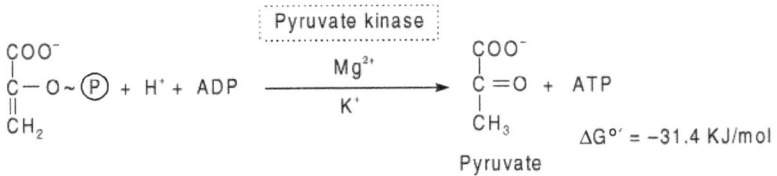

$$
\underset{\text{Pyruvate}}{\begin{array}{c}
\text{COO}^- \\
|\\
\text{C}-\text{O}\sim \text{(P)} + \text{H}^+ + \text{ADP} \\
\|\\
\text{CH}_2
\end{array}}
\quad
\xrightarrow[\text{K}^+]{\overset{\text{Pyruvate kinase}}{\text{Mg}^{2+}}}
\quad
\underset{\text{Pyruvate}}{\begin{array}{c}
\text{COO}^- \\
|\\
\text{C}=\text{O} + \text{ATP} \\
|\\
\text{CH}_3
\end{array}}
$$

$\Delta G^{\circ\prime} = -31.4 \text{ KJ/mol}$

The enzyme pyruvate kinase catalyses the transfer of phosphate group from phosphoenol pyruvate to ADP in another substrate-level phosphorylation. The enzyme requires Mg^{2+} and K^+. The reaction is highly exergonic.

The overall reaction is as follows:

Glucose $+ 2ADP + 2Pi + 2NAD^+ \longrightarrow 2$ pyruvate $+ 2ATP + 2NADH^+ + 2H^+ + 2H_2O$

The overall reactions in glycolysis are shown in figure 9.1.

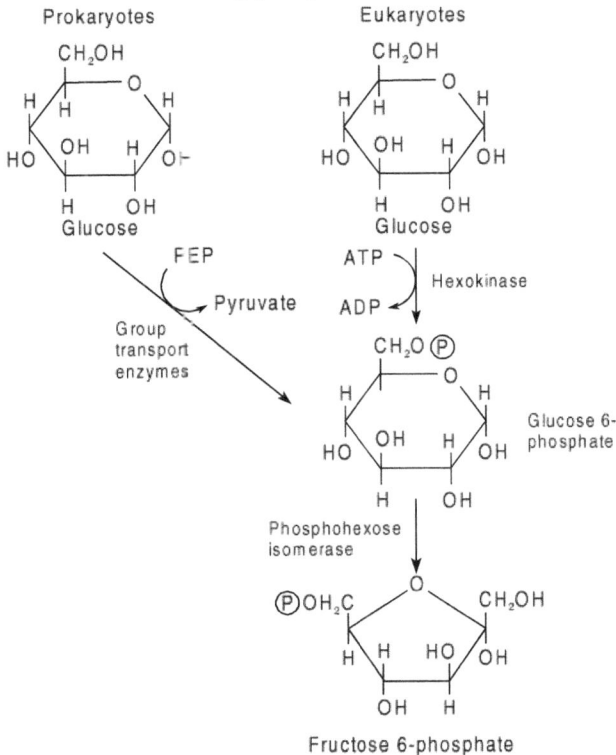

Fructose 6-phosphate

(Contd.)

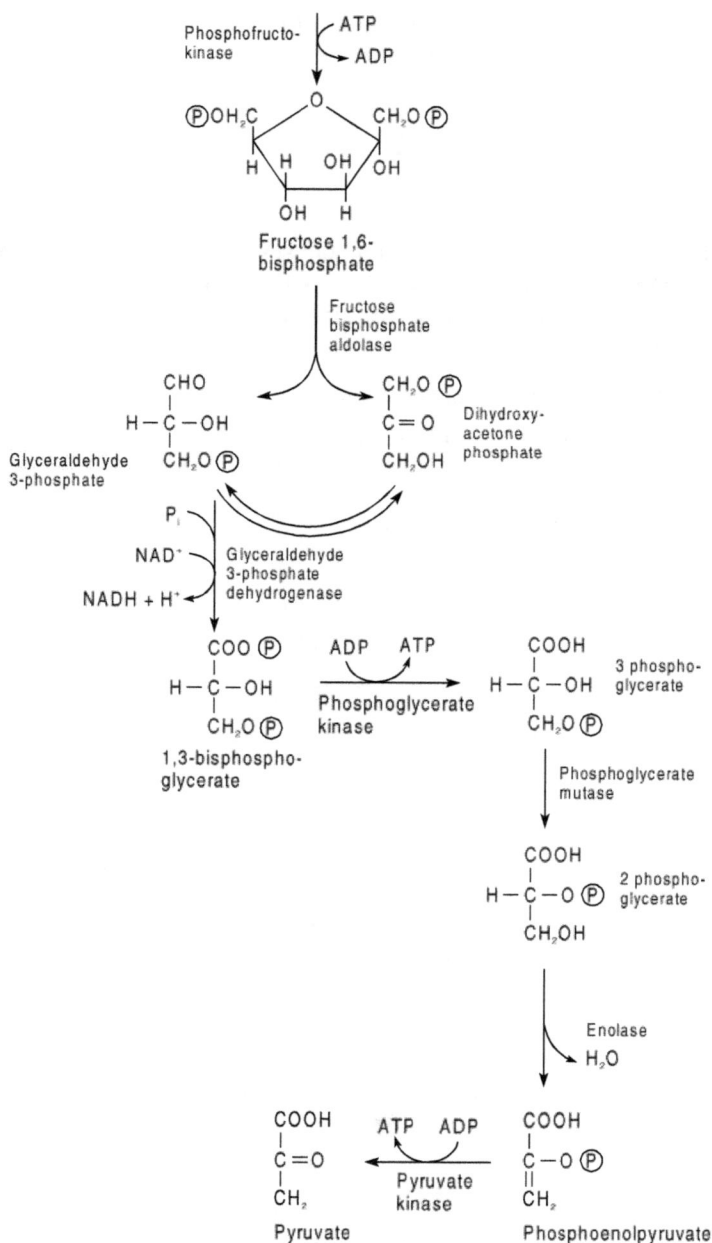

Figure 9.1   Overall reactions in glycolysis.

STUDY OUTLINE

- Glycolysis is a central metabolic pathway that is found throughout nature.

- Glycolysis is the pathway through which the longest flux of carbon occurs in most cells.

- Glycolysis consist of two stages (i) preparatory phase (or) energy investment phase and (ii) redox reaction (or) energy generating phase.

- The preparatory phase is called energy investment phase because two ATP molecules are consumed in this phase.

- The hexose sugar glucose is metabolized and it gives pyruvate which synthesizes 8 ATP molecules.

CONCEPT CHECK

1. Explain the preparatory phase and redox reactions.

2. How many ATP molecules are synthesized during the oxidation of one molecule of glucose?

10

CITRIC ACID CYCLE, TRICARBOXYLIC ACID CYCLE OR KREB'S CYCLE

INTRODUCTION

The early steps in the respiration of glucose involve the same biochemical steps as those of glycolysis. A key intermediate in glycolysis is pyruvate — while in fermentation, pyruvate is converted to the fermented products, in respiration, pyruvate is oxidized fully to CO_2. One major pathway by which pyruvate is completely oxidized to CO_2 is called the citric acid cycle. In other words it refers to the molecular processes involved in O_2 consumption and CO_2 formation by cells. This is called as cellular respiration.

Generally cellular respiration occurs in 3 stages.

I Stage

In the first stage organic fuel molecules like glucose, fatty acids, and some amino acids are oxidized to yield 2-carbon fragments in the form of the acetyl group of acetyl coenzyme A (acetyl-CoA).

II Stage

In the second stage, these acetyl groups are fed into the citric acid cycle which enzymatically oxidizes them to CO_2. The energy released by oxidation is conserved in the reduced electron carriers, NADH and $FADH_2$, of respiration.

III Stage

These reduced cofactors are themselves oxidized, giving up protons (H^+) and electrons. These electrons are transferred along a chain of electron-carrying molecules known as the respiratory chain, to O_2, which they reduce to form H_2O. During this process of electron transfer, much energy is released and conserved in the form of ATP, in the process called oxidative phosphorylation. The citric acid cycle is therefore an amphibolic pathway that operates both catabolically and anabolically. Kreb's cycle marks the "hub" of the metabolic system. It accounts for the major portion of carbohydrate, fatty acid, and amino acid oxidation and generates numerous biosynthetic precursors. In aerobic organisms, glucose and other sugars, fatty acids and most of the amino acids are ultimately oxidized to CO_2 and H_2O via the citric acid cycle. Before they can enter the cycle, the carbon skeletons of sugars and fatty acids must be degraded to the acetyl group of acetyl CoA, the form in which the citric acid cycle accepts most of the fuel input.

CONVERSION OF PYRUVATE TO ACETATE

The conversion of pyruvate to acetyl CoA and CO_2 is by a structured cluster of three enzymes, "the pyruvate dehydrogenase complex," located in the mitochondria of eukaryotic cells and in the cytosol of prokaryotes.

Step I

Pyruvate is converted into acetyl CoA by three enzyme complexes and one $NADH_2$ is released by oxidative decarboxylation. In addition to this, five cofactors are also needed for the reaction. They are as follows:

Thiamine Pyrophosphate (TPP) catalyses decarboxylation of pyruvate to acetaldehyde in alcoholic fermentation.

Pantothenate is essential for component of coenzyme.

Coenzyme A has a reactive thiol (–SH) group that is critical to the role of acyl carriers. In a number of metabolic reactions acyl group becomes covalently linked to the thio group and forms thioesters which have a high acyl group transfer potential donating their acyl group to a variety of acceptors.

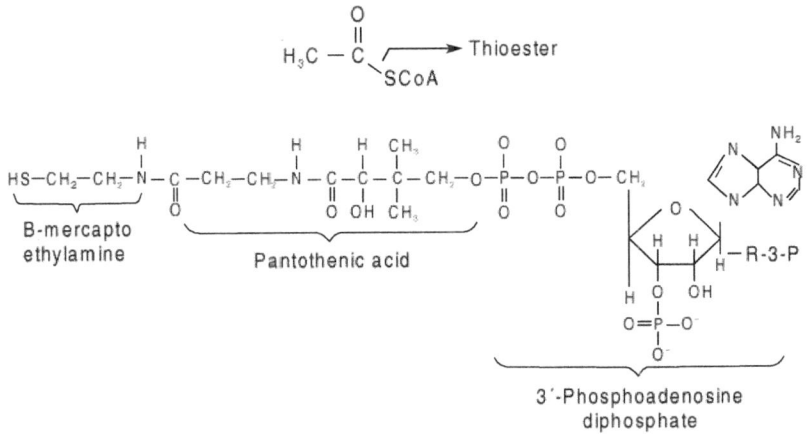

FAD and NAD[+] act as electron carriers. Lipoate has two thiol groups, both essential to its role as cofactors. In reduced state, both sulphur atoms are present as –SH group, but oxidation of this sulphur atoms produces (–S–S–) bond similar to that between two cysteine residues

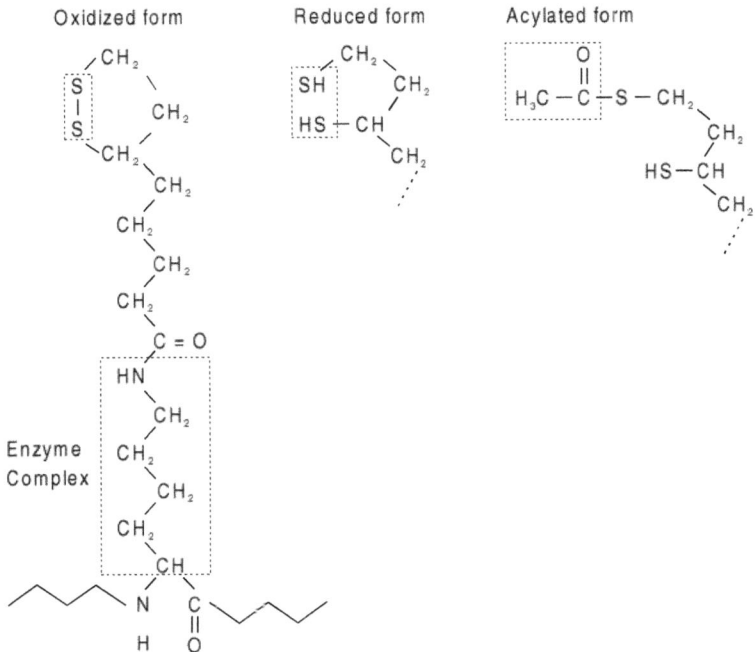

Figure 10.1 Various forms of lipoate.

in a protein. Because of this capacity to undergo oxidation–reduction reaction, lipoate can serve both as an electron carrier and as an acyl carrier. Both the functions are important in the action of pyruvate dehydrogenase.

Pyruvate Dehydrogenase Complex

Pyruvate dehydrogenase consists of multiple copies of three enzymes.

- Pyruvate dehydrogenase (E_1)
- Dihydrolipoyl transacetylase (E_2)
- Dihydrolipoyl dehydrogenase (E_3)

Before entering into the kreb's cycle, the pyruvate is converted into acetyl-CoA with the help of the enzyme pyruvate dehydrogenase.

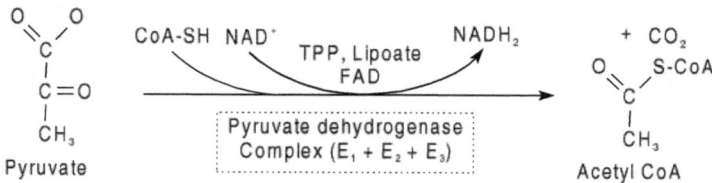

Generally the citric acid cycle is carried out in 8 steps. The steps are as follows.

REACTION 1

Formation of citrate

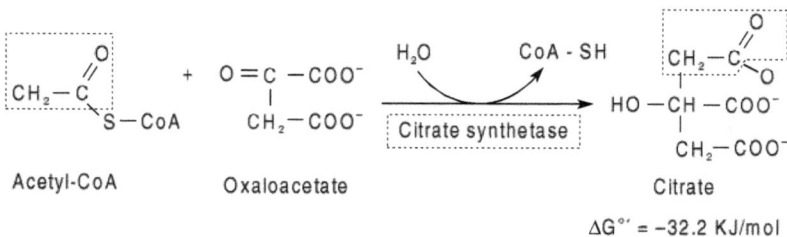

$$\Delta G^{\circ\prime} = -32.2 \text{ KJ/mol}$$

In reaction 1, the condensation of acetyl-CoA with oxaloacetate to form citrate is catalysed by citrate synthetase. This reaction is highly exergonic. In this reaction the methyl carbon of the acetyl group is joined to the carbonyl group (C–2) of oxaloacetate.

REACTION 2

Formation of isocitrate via cis-aconitase

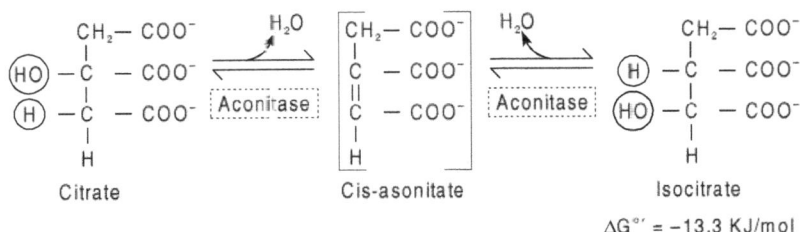

Citrate   Cis-aconitate   Isocitrate

$$\Delta G^{\circ\prime} = -13.3 \text{ KJ/mol}$$

The enzyme aconitase catalyses the formation of isocitrate from citrate, through the intermediary formation of the tricarboxylic acid cis-aconitate. Aconitase contains an iron-sulphur centre, which acts both in binding of the substrate at the active site and in the addition or removal of H_2O.

REACTION 3

Oxidation of isocitrate to α-ketoglutarate and CO_2.

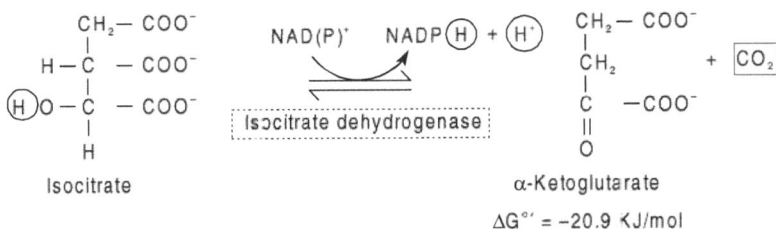

Isocitrate α-Ketoglutarate

$$\Delta G^{\circ\prime} = -20.9 \text{ KJ/mol}$$

In reaction 3, the substrate isocitrate is converted into α-ketoglutarate with the help of the enzyme isocitrate dehydrogenase. This reaction is exergonic. The product formed from this reaction acts as substrate for various metabolic pathways. So α-ketoglutarate is an important intermediate in citric acid cycle.

REACTION 4

Oxidation of α-ketoglutarate to succinyl-CoA and CO_2

$$
\begin{array}{l}
\text{CH}_2-\text{COO}^- \\
| \\
\text{CH}_2-\text{COO}^- \\
\| \\
\text{O}
\end{array}
\quad
\xrightarrow[\text{dehydrogenase complex}]{\text{CoA-SH} \quad \text{NAD}^+ \quad\quad\quad \text{NADH}}
\quad
\begin{array}{l}
\text{CH}_2-\text{COO}^- \\
| \\
\text{CH}_2 \quad\quad + \text{CO}_2 \\
| \\
\text{C}-\text{SCoA} \\
\| \\
\text{O}
\end{array}
$$

α-Ketoglutarate   Succinyl-CoA

$\Delta G^{\circ\prime} = -33.5$ KJ/mol

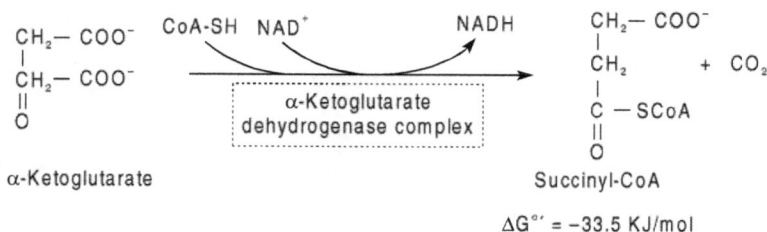

The energy is produced with the loss of CO_2 during the oxidation of α-ketoglutarate. This reaction is highly exergonic. The succinyl-CoA formation from α-ketoglutarate is mediated by the enzyme α-ketoglutarate dehydrogenase complex. The energy during oxidation is conserved in the thioester bond of succinyl-CoA. α-ketoglutarate dehydrogenase consists of three enzymes which are analogues to E_1, E_2 and E_3 of the pyruvate dehydrogenase complex, as well as enzyme-bound TPP, bound lipoate, FAD, NAD and coenzyme A.

REACTION 5

Conversion of succinyl-CoA to succinate

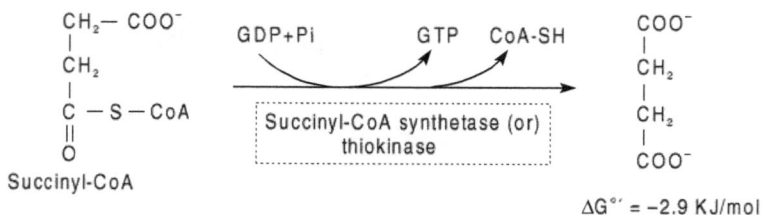

$$
\begin{array}{l}
\text{CH}_2-\text{COO}^- \\
| \\
\text{CH}_2 \\
| \\
\text{C}-\text{S}-\text{CoA} \\
\| \\
\text{O}
\end{array}
\quad
\xrightarrow[\text{thiokinase}]{\text{GDP+Pi} \quad\quad \text{GTP} \quad \text{CoA-SH}}
\quad
\begin{array}{l}
\text{COO}^- \\
| \\
\text{CH}_2 \\
| \\
\text{CH}_2 \\
| \\
\text{COO}^-
\end{array}
$$

Succinyl-CoA   Succinyl-CoA synthetase (or)

$\Delta G^{\circ\prime} = -2.9$ KJ/mol

Both the enzyme names succinyl-CoA synthetase and thiokinase indicate the participation of the nucleoside triphosphate in the reaction. The enzyme mediates the conversion of succinyl-CoA to succinate. In the previous reaction, α-ketoglutarate conversion of succinyl-CoA releases energy and is conserved in the thioester bond of succinyl-CoA. But in reaction 6, the conserved energy in thioester is released and is utilized for the production of ATP from ADP. This is called as substrate level phosphorylation. In the case of prokaryotes, ADP is utilized for this reaction. But in eukaryotes GDP is utilized and released as GTP whereas in prokaryotes it is released as ATP.

REACTION 6

Oxidation of succinate to fumarate

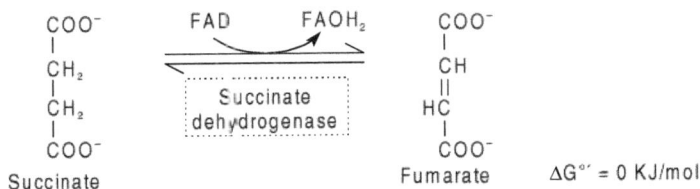

$$
\begin{array}{ccc}
COO^- & \xrightarrow[\text{Succinate dehydrogenase}]{\text{FAD} \quad \text{FAOH}_2} & COO^- \\
| & & | \\
CH_2 & & CH \\
| & & \| \\
CH_2 & & HC \\
| & & | \\
COO^- & & COO^-
\end{array}
$$

Succinate   Fumarate $\Delta G^{o\prime} = 0$ KJ/mol

The fumarate is formed by donating two protons to FAD from succinate. This reaction is mediated by the enzyme succinate dehydrogenase. The reaction is completed with the release of $NADH_2$. The enzyme succinate dehydrogenase is the only membrane-bound enzyme of the citric acid cycle. In eukaryotes it is bound to the inner mitochondrial membrane. In prokaryotes it is bound to the plasma membrane.

REACTION 7

Hydration of fumarate to produce malate

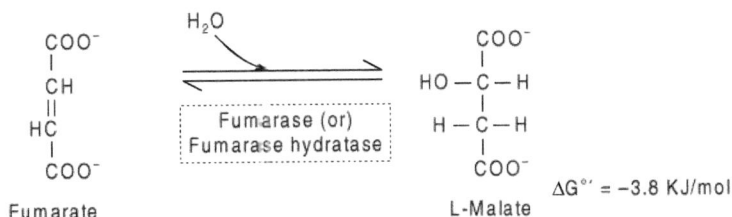

$$
\begin{array}{ccc}
COO^- & \xrightarrow[\text{Fumarase (or) Fumarase hydratase}]{H_2O} & COO^- \\
| & & | \\
CH & & HO-C-H \\
\| & & | \\
HC & & H-C-H \\
| & & | \\
COO^- & & COO^-
\end{array}
$$

$\Delta G^{o\prime} = -3.8$ KJ/mol

Fumarate   L-Malate

This reaction is slightly exergonic. This reaction is catalysed by the enzyme fumarase or fumarase hydratase. This enzyme mediates this reaction with the help of H_2O. The substrate is hydrated. The enzyme fumarase is highly stereospecific; it catalyses the hydration of trans double bond of fumarate but does not act on malate, the isomer of fumarate. In the reverse direction (from malate to fumarate), fumarase is equally stereospecific. But D-malate cannot act as a substrate.

REACTION 8

Oxidation of malate to oxaloacetate

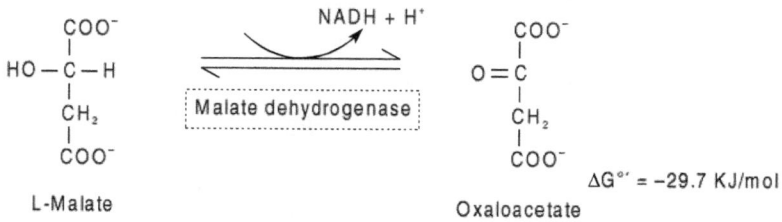

$\Delta G^{\circ\prime} = -29.7$ KJ/mol

L-Malate   Oxaloacetate

Malate dehydrogenase catalyses the oxidation of L-malate to oxalo acetate. However, in intact cells oxaloacetate is continually removed

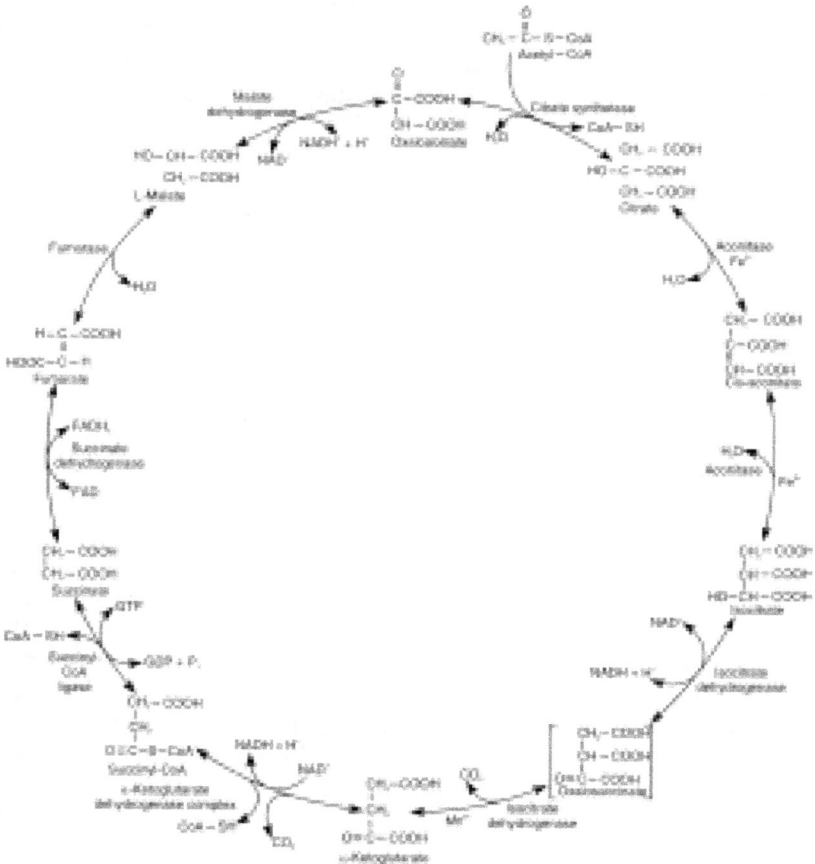

Figure 10.2   The overall reaction in kreb's cycle.

by the highly exergonic citrate synthetase reaction. This keeps the concentration of oxaloacetate in the cell extremely low (10^{-6}M) pulling the malate dehydrogenase reaction toward oxaloacetate formation.

STUDY OUTLINE

- The complete oxidation of pyruvate to CO_2 is termed as citric acid cycle.

- Citric acid cycle is otherwise called as cellular respiration.

- Totally 30 ATP molecules are synthesized during citric acid cycle.

- Various enzymes and high-energy compounds are involved in this cycle.

- This cycle is otherwise called as tricarboxylic acid cycle or kreb's cycle.

- The first step in kreb's cycle is the conversion of pyruvate to acetyl CoA and CO_2 by a structured clusters of three enzymes "the pyruvate dehydrogenase complex" located in the mitochondria of eukaryotes and in the cytosol of prokaryotes.

CONCEPT CHECK

1. Explain about the energetics of TCA cycle.
2. Outline the steps of citric acid cycle.

CRITICAL THINKING

How many ATPs are produced during TCA cycle? Write the energetics of citric acid cycle.

11

HEXOSE MONOPHOSPHATE
PATHWAY (HMP SHUNT)

In most microbes, the EM pathway is the major pathway for glucose conversion to pyruvate. But the EM pathway fails to provide the five sugars required for DNA and RNA formation. It also fails to provide erythrose 4-phosphate, a critical intermediate for aromatic amino acid synthesis. Hexose monophosphate (HM) pathway provides all these materials. The hexose monophosphate shunt (HMS) pathway, also known as the oxidative pentose (OP) pathway and variations of it are widely found in microbes. The pathway starts with the conversion of glucose 6-phosphate formed by the action of hexokinase on glucose, to 6-phosphogluconogamma lactone, a process mediated by glucose 6-phosphate dehydrogenase that also allows conversion of NADP to NADPH. After this step, the lactone is hydrolysed with lactonase, yielding 6-phosphogluconic acid dehydrogenase in the presence of NADP, yielding NADPH, carbon dioxide (CO_2), and through ribulose-5-phosphate, a mixture of ribose 5-phosphate and xylulose 5-phosphate. When the "complete" pathway occurs, a molecule of xylulose 5-phosphate transfers two carbons in a transketolase-mediated reaction to yield the seven carbon molecules, sedoheptulose 7-phosphate, which arises from the top two carbons of a xylulose 5-phosphate and the five carbons of ribulose 5-phosphate molecule. Glyceraldehyde 3-phosphate is also formed from the bottom three carbons of the xylulose 5-phosphate molecule.

In the next step, sedoheptulose-7-phosphate interacts with glycerdehyde 3-phosphate and forms fructose 6-phosphate and a molecule of erythrose 6-phosphate which arises from the bottom of four carbons of sedoheptulose molecule. At this point, the erythrose 4-phosphate reacts with the second molecule of xylulose 5-phosphate through a transketolase reaction to yield a second molecule of fructose 6-phosphate, in which the bottom four carbons are derived from erythrose 4-phosphate and the top 2 carbons originate from xylulose 5-phosphate. Then the third molecule of fructose 6-phosphate is formed through the condensation of two glyceraldehyde 3-phosphate molecules by a reversal of aldolase and subsequent conversion of fructose 1,6-bisphosphate through the enzyme phosphorylase to fructose 6-phosphate. Isomerization of the fructose 6-phosphate to glucose 6-phosphate allows the reinitiation of the cycle. After completion of the cycle, the products are CO_2, NADPH, and the intermediates for biosynthesis, such as erythrose 4-phosphate and ribose 5-phosphate. If the sequence of reactions operate completely, hexose molecules are converted to six pentoses and six CO_2

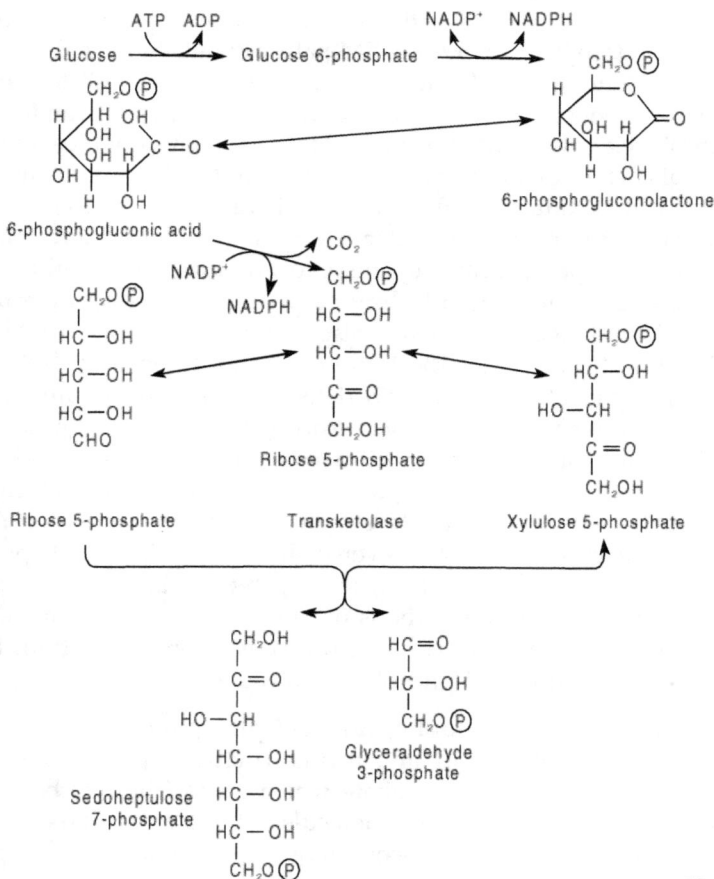

(Contd.)

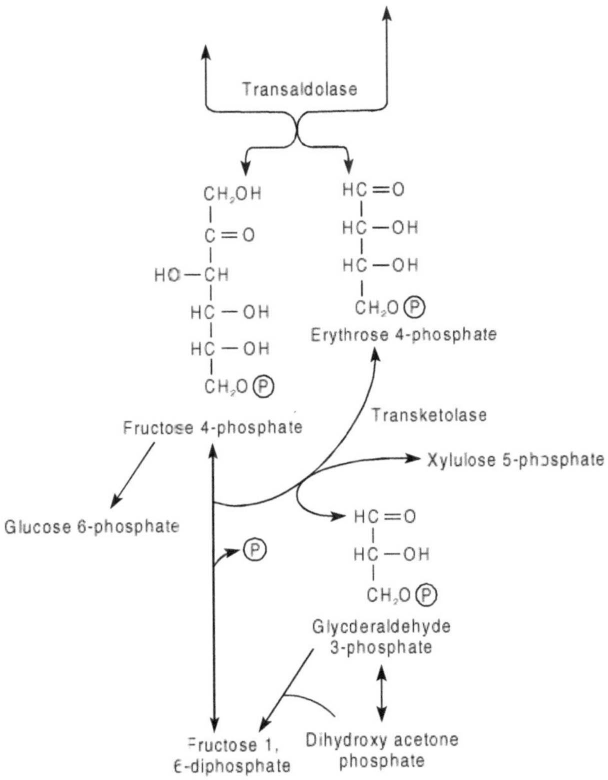

Net reaction:

Glucose + 6 NADP$^+$ → Glyceraldehyde 3-phosphate + 3CO$_2$ + 6NADPH

Figure 11.1 The pentose phosphate pathway of glucose oxidation. Precursor metabolites are shown.

molecules, and the six pentose molecules are reconverted to hexoses. No ATP formation is possible from its operation, unless it results from conversion of NADPH to NADH by transhydrogenase reaction and the use of NADH in electron transport phosphorylation.

THE ENTNER–DOUDOROFF PATHWAY

This pathway is also considered as an alternate hexose monophosphate (HM) pathway to the oxidative pentose pathway. This pathway provides the five critical biosynthetic intermediates namely glucose 6-phosphate, triose phosphate, 3-phosphoglycerate, phosphoenol

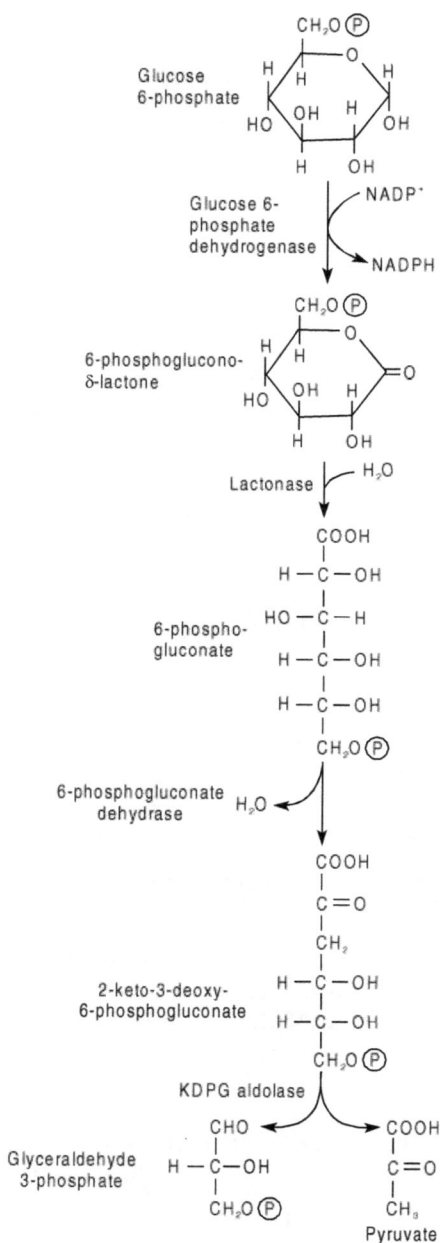

Figure 11.2 The Entner–doudoroff pathway.

pyruvate, and pyruvate. The reaction begins with glucose which is converted to glucose 6-phosphate mediated by the enzyme hexokinase. This is further metabolized by the enzyme glucose 6-phosphate dehydrogenase and lactonase to 6-phosphogluconic acid. Rather than being converted to pentoses and carbon dioxide, it is dehydrated through 6-phosphogluconic acid dehydratase to yield 2-keto,3-dehydro 6-phosphogluconic acid. Under the influence of 3-keto 3-dehydro 6-phosphogluconic acid aldolase, the acid gives rise to pyruvate directly from the top three hexose carbons and glyceraldehyde 3-phosphate from the bottom three glucose carbons. The glyceraldehyde 3-phosphate is further converted through EM pathway reactions to pyruvate.

STUDY OUTLINE

- Hexose monophosphate pathway, also known as the oxidative pentose pathway (OP), provides the five sugars required for DNA and RNA formation and also provides erythrose 4-phosphate, a critical intermediate for aromatic amino acid synthesis.

- Entner–Doudoroff pathway is otherwise the alternate hexose monophosphate (HM) pathway to oxidative pentose pathway.

- The five critical biosynthetic intermediates are glucose 6-phosphate, triose phosphate, 3-phosphoglycerate, phosphoenol pyruvate, and pyruvate.

CONCEPT CHECK

1. Explain the steps involved in hexose monophosphate pathway.
2. Briefly explain about the ED pathway.

CRITICAL THINKING

In this chapter it is stated that AMP pathway provides the precursor for amino acid and nucleic acid biosyntheses. But not all the organisms undergo this type of alternative pathway. In that case how do the organisms derive their precursors.

12

CARBOHYDRATE BIOSYNTHESIS

INTRODUCTION

Carbohydrates are major cell constituents and are the building blocks of many internal organelles of a cell. They play an important role in cell wall of microorganism. From simple substrates, cells synthesize the various types of macromolecules needed for the cells to live and grow. The substrate for each cell constituent may be the inorganic compounds such as CO_2 used by autotrophs or the organic compounds such as glucose used by heterotrophs. The major compounds formed from these substrates include proteins (which act as enzymes, structural proteins, membrane carriers, receptors and so on), lipids (which serve as the main components of membranes), carbohydrates (which form much of the structure of cell walls) and nucleic acids (which store and express genetic information). All these compounds are composed of nitrogen, carbon, oxygen and hydrogen.

GLUCONEOGENESIS

Gluconeogenesis can be defined as the biosynthesis of glucose from noncarbohydrate compounds. Glucose biosynthesis is an important event in the metabolism of a cell because it is a carbohydrate which acts as the major constituent of cell wall, cell membrane and, nucleic acids like DNA and RNA and as a glycoprotein.

Once glucose is formed, it enters into pentose phosphate pathway to supply 5-carbon carbohydrate molecules such as ribose and deoxyribose sugars needed for the synthesis of nucleic acids.

In the gluconeogenesis pathway, the substrate is converted into pyruvate or another intermediate in the pathway, which is then converted to glucose. Amino acids derived from proteins can be converted into pyruvate or phosphoenol pyruvate, which are intermediary metabolites of the entire gluconeogenic pathway. In some other way the substrate can be obtained for gluconeogenic pathway by the breakdown of lipids into the 3-carbon intermediates. Glycolysis and gluconeogenesis pathway have same intermediates but the pathways are actually different. The enzymes involved in this pathway are also different. In each direction there is a critical enzymatic step that is irreversible, meaning that the enzyme concerned catalyses the reaction in one direction only. For example, during the Embden–Meyerhof pathway of glycolysis, the conversion of fructose 6-phosphate to fructose 1,6-bisphosphate is catalysed irreversibly by the enzyme phosphofructokinase. But in the case of gluconeogenesis, the conversion of fructose 1,6-bisphosphate to fructose 6-phosphate is catalysed irresversibly by the enzyme fructose 1,6-bisphosphatase. These enzymes have different allosteric inhibitors that regulate the rate of carbon flow in either direction. The synthesis of carbohydrates is favoured when the cells have an adequate supply of ATP. The catabolism is favoured when ATP concentrations are relatively low.

Table 12.1   Some key reactions in the glycolytic and glyconeogenic pathways in the enzyme that catalyses the reaction.

Glycolysis reaction (Enzyme)	Gluconeogenesis Reaction (Enzyme)
Glucose → Glucose 6-phosphate (*Hexokinase*)	Glucose 6-phosphate → glucose (*Glucose 6-phosphatase*)
Fructose 6-phosphate → fructose 1, 6-bisphosphate (*Phosphofructokinase*)	Fructose 1, 6-bisphosphate → fructose 6-phosphate (*Fructose 1,6- bisphosphatase*)
Phosphoenol pyruvate → pyruvate (*Pyruvate kinase*)	Pyruvate → phosphoenol pyruvate (*Pyruvate carboxylase/ phosphoenol pyruvate carboxy kinase*)

Glucose

↑ Glucose 6-phosphatase

Glucose 6-phosphate

↑

Fructose 6-phosphate

↑ Fructose 1,6-bisphosphatase

Fructose 1,6-bisphosphate

Glyceraldehyde 3-phosphate ⟷ Dihyroxy phosphate

↑

1,3-Bisphosphoglycerate

↑ ADP / ATP

3-Phosphoglycerate

↑

2-Phosphoglycerate

↑

Phosphoenolpyruvate

↑ Phosphoenolpyruvate carboxy kinase → CO_2 → GDP / GTP

Oxaloacetate

↑ Pyruvate carboxylase ⟷ Amino acids + Fatty acids

Pyruvate

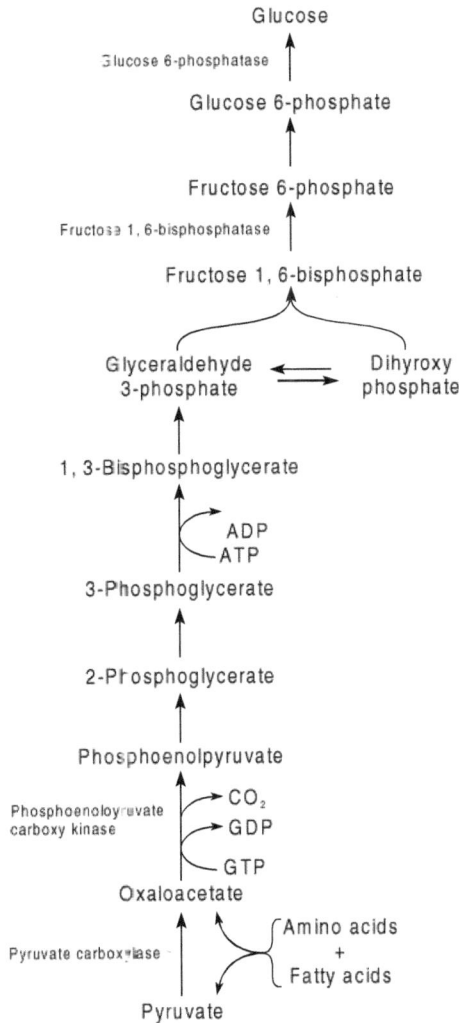

Figure 12.1 Gluconeogenic pathway.

GLYOXYLATE PATHWAY

Some microorganisms use this type of pathway. Glyoxylate pathway permits the flow of carbon from fatty acids (lipids) or acetate to carbohydrates. This pathway is a shunt or "short circuit" across the tricarboxylic acid cycle that serves to replenish oxaloacetate in the cell. This type of pathway is important because, the key intermediate of the tricarboxylic acid cycle may be used for the biosynthesis of other organic molecules. Reactions in a cell that serve to replenish the supplies of key molecules are called anaplerotic sequences (which means "filling up"). In the glyoxylate cycle, isocitrate is split by isocitrate lyase into succinate and glyoxylate. Malate is formed from the reaction of glyoxylate with acetyl-CoA. The formed malate is converted through oxloacetate to phophoenol pyruvate, an intermediate metabolite of the glyconeogenic pathway. This links the pathway of fatty acid metabolism with the pathway of carbohydrate metabolism, allowing four molecules of acetyl-CoA to participate in the formation of glucose.

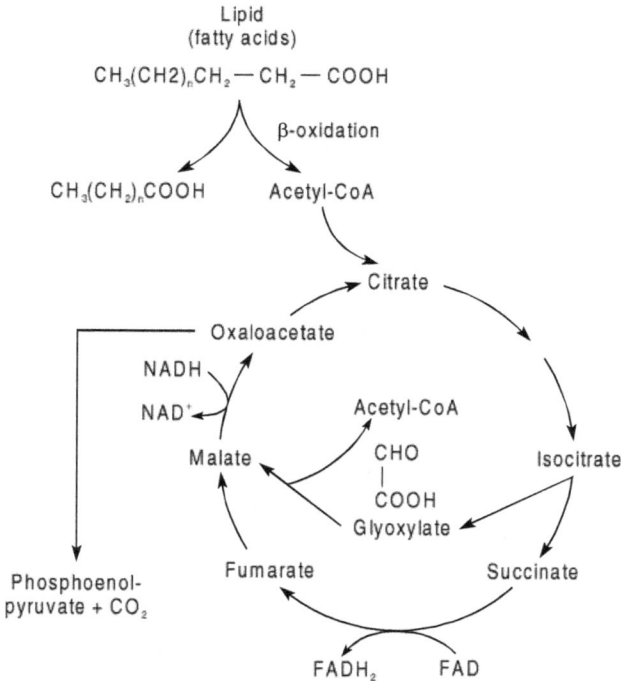

Figure. 12.2 Glyoxylate cycle.

The glyoxylate cycle is a shunt across the normal tricarboxylic acid cycle. This pathway is needed to maintain the tricarboxylic acid cycle intermediates when lipids are metabolized. It also provides a way of converting lipids into carbohydrates.

The reactions are as follows

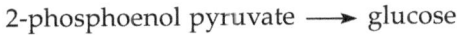

$$4 \text{ acetyl-CoA} \longrightarrow 2 \text{ oxloacetate} \longrightarrow 2\text{-phosphoenol pyruvate} + 2CO_2$$

$$2\text{-phosphoenol pyruvate} \longrightarrow \text{glucose}$$

LIPOPOLYSACCHARIDE BIOSYNTHESIS (LPS)

Lipopolysaccharide is an important constituent of the cell wall. It is mainly present in the gram-negative cell wall. The negative charge of the gram-negative cell wall is mainly contributed by the lipopolysaccharide. LPS acts as antigen and is responsible for many diseases. LPS also act as attachment site or receptor for many bacteriophages. The biosynthesis of lipopolysaccharides and its addition to the gram-negative cell wall occurs at the cytoplasmic membrane in successive steps. For the synthesis of peptidoglycan, undecaprenyl phosphate or bactoprenol serves as the lipid carrier to

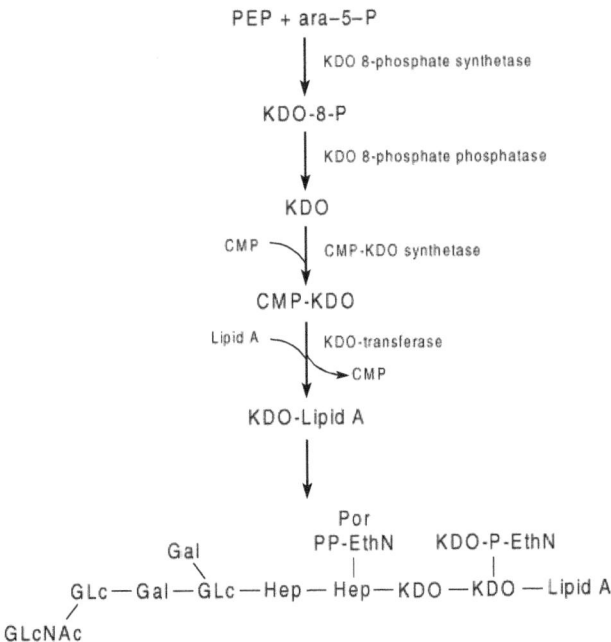

PEP + ara–5–P

↓ KDO 8-phosphate synthetase

KDO-8-P

↓ KDO 8-phosphate phosphatase

KDO

CMP ⟍ | CMP-KDO synthetase

CMP-KDO

Lipid A ⟍ | KDO-transferase

➤ CMP

KDO-Lipid A

↓

Por
PP-EthN   KDO-P-EthN
Gal | |
\   GLc—Gal—GLc—Hep— Hep—KDO —KDO —Lipid A
/
GLcNAc

Figure. 12.3 LPS biosynthesis.

which individual sugars of the repeat unit (the outermost portion of the LPS molecule) are sequentially added. These sugars are first activated by condensing with nucleotide triphosphates to form nucleotide diphosphate—sugar intermediates such as GDP-mannose, UDP-galactose, or CDP-abequose. The activated sugars are then enzymatically added in a particular sequence, thus forming a polysaccharide chain of repeating sugar units. The lipid A fraction is assembled in the cytoplasmic membrane by condensing two molecules of glucosamine phosphate and then adding several fatty acid molecules, particularly β hydroxymyristic acid. The core polysaccharide is built on the lipid fraction by the sequential enzymatic addition of core sugars such as KDO (2-keto 3-deoxy octulosonic acid), heptose, galactose and glucose. The lipid A core polysaccharide is then translocated to the outer surface of the cytoplasmic membrane, after which the repeat polysaccharide is transferred from the undecaprenyl phosphate carrier to the lipid A core polysaccharide, thus completing the synthesis of lipopolysaccharide.

The synthesis of core polysaccharide linked to lipid A is an essential part of LPS biosynthesis. Ketodeoxyoctulosonic acid (KDO) is formed by the reaction of phosphoenolpyruvate (PEP) and arabinose 5-phosphate (ara 5-P). Cytosine monophosphate (CMP) is added to KDO during the pathway and later removed when KDO is attached to lipid A. Two subsequent additions of KDO and the inclusion of several other sugars, including heptose (Hep), glucose (Glc), N-acetylglucosamine (GlcNAc), galactose (gal), and ethanolamine (EthN) are also added.

PEPTIDOGLYCAN BIOSYNTHESIS

Biosynthesis of peptidoglycan is essential for cell division and growth. Peptidoglycan is a major component of cell wall of microorganisms. Based on the content of peptidoglycan, eubacteria are classified into gram-positive and gram-negative microorganisms. Gram-positive organisms have high amount of peptidoglycan than gram-negative organisms. The presence of peptidoglycan protects the cells from solvent damage and provides a rigid shape to the organism. They are also involved in the transport mechanism. Peptidoglycan acts as a basal plate for flagella attachment. The glycan portion of the bacterial cell wall is a disaccharide composed of acetyl glucosamine and acetyl muramic acid. The formation of acetyl glucosamine involves the conversion of glucose to fructose 6-phosphate and subsequent reaction with glutamic acid and acetyl CoA. For the formation of the second component (N-acetyl muramic acid), N-acetylglucosamine reacts with uridine triphosphate (UTP) to form acetyl glucosamine uridine diphosphate (UDP), which then reacts with phosphoenol pyruvate to form N-acetyl muramic acid UDP. During the first stage of cell wall synthesis, the precursors of peptidoglycan are assembled in the cytoplasm to form a UDP N-acetyl muramic acid pentapeptide. The pentapeptide is composed of a tetrapeptide that occurs in the cell wall with the additional D-alanine at the carboxyl end of the chain. In some bacteria like *Escherichia coli* and *Bacillus subtilis,* the pentapeptide would be UDP N-acetyl muramic acid-L-alanine-D-glutamate-diaminopimelic acid-D-alanine. The structure is linked by a peptide bond, and synthesized by enzymes in the cytoplasm rather than on ribosomes where most other peptide bonds are formed. Each amino acid is added to the appropriate place by specific adding enzymes or ligases whose activity requires ATP. D-alanine is formed from L-alanine by alanine racimase and then two D-alanine molecules are connected to form a D-alanine-D-alanine dipeptide by D-alanine-D-alanine synthetase.

In stage II, the N-acetyl muramic acid pentapeptide is transferred to a carrier molecule in the second stage of peptidoglycan biosynthesis. The carrier is called as C55 carrier lipid, undecaprenyl phosphate, or bactoprenol. It is located with the cytoplasmic membrane where the second stage of cell wall biosynthesis occurs. The C55 carrier exists in the cell as a pyrophosphate (PPi) that must be dephosphorylated by a pyrophosphate before it can act as a carrier. The N-acetyl glucosamine is transferred to the bactoprenol-N-acetyl muramic acid pentapeptide to form bactoprenol-N-acetyl glucosamine-N-acetyl muramic acid penta peptide. This molecule moves across the cytoplasmic membrane (translocation) to the outside of the cell. There, it is transferred to a growing chain of cell wall precursors with the cell wall called "nascent peptidoglycan."

In the final stage of peptidoglycan biosynthesis, the nascent peptide glycan is covalently bound to the existing cell wall by transpeptidation or by the formation of cross bridges between existing cell wall peptidoglycan and those on the nascent peptidoglycan. In this reaction, the terminal D-alanine of the pentapeptide is cleared and the energy released by this is used to attach the fourth amino acid (D-alanine) of the remaining tetrapeptide to diaminopimelic acid of an adjacent tetrapeptide, which is already part of the wall. Thus the nascent or

Stage I
(cytoplasm)

N-acetylglucosamine

— UTP

→ PP,

N-acetylglucosamine-UDP
(G-UDP)

— Phosphoenolpyruvate

N-acetylmuramic acid-UDP
(M-UDP)

L-ala ———— ATP

D-glu ———— ATP

L-ala

↓

D-ala

L-lys ———— ATP

D-ala
|
D-ala ———— ATP

M-UDP
|
L-ala
|
D-glu   M-UDP
| |
L-lys   Pentapeptide (pp)
|
M-UDP   D-ala
|
PP   D-ala

Stage II
(cytoplasmic   M-carrier lipid
membrane) |
                    L-ala ———— M-carrier lipid
                    | |
                    D-glu   PP
                    |
                    DAP   — G-UDP
                    |
                    D-ala
                    | G-M-carrier lipid
                    D-ala |
                               PP

Stage III
(cell wall)

Linkage of glycan
+
Cross-linkage of Peptide

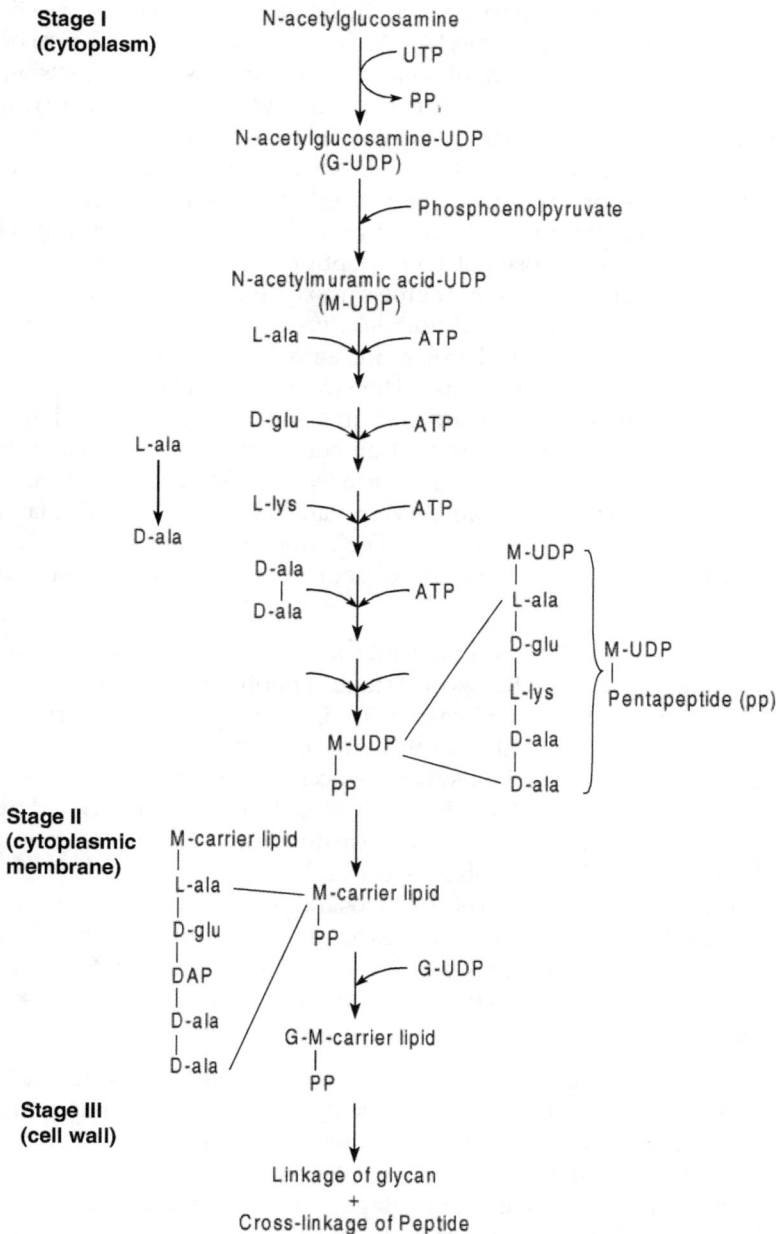

Figure 12.4   Peptidoglycan biosynthesis.

newly formed peptidoglycan chains become added to the existing cell wall. Several enzymes are involved in the peptidoglycan biosynthesis, which bind to pencillin and are called as pencillin-binding proteins (PBPs). Binding of pencillin to PBS inhibits their activities.

The synthesis of peptidoglycan by bacteria is important to provide the backbone material of the cell wall. Uridine triphosphate (UTP) activates N-acetyglucosamine and reacts with phosphoenolpyruvate to form N-acetylmuramic acid-UDP. To form peptidoglycan, amino acids must be added to N-acetylmuramic acid, a repeating and alternating glycan chain of N-acetylmuramic acid and N-acetylglucosamine must be formed, and the peptide chains must be cross linked.

STUDY OUTLINE

- Carbohydrates are a major cell constituent and the building blocks of many internal organelles of a cell.

- Gluconeogenesis can be defined as the biosynthesis of glucose from noncarbohydrate compounds. Glucose biosynthesis is an important event in the metabolism of cell.

- In the case of glyoxylate pathway, it permits the flow of carbon from acids (lipids) or acetate to carbohydrates. This pathway is a shunt or short circuit across the tricarboxylic cycle that serves to replenish oxalate in the cell.

- Lipopolysaccharide is an important constituent of the cell wall. The biosynthesis of lipopolysaccharides and its addition to the gram-negative cell wall occur at the cytoplasmic membrane.

- Peptidoglycan biosynthesis is essential for cell division and growth. Peptidoglycan is a major component of the cell wall of microorganisms.

CONCEPT CHECK

1. How is peptidoglycan synthesized?
2. Explain the process of biosynthesis of lipopolysaccharide.
3. What is gluconeogenesis? Explain the steps in gluconeogenesis.
4. What is glyoxylate pathway? Explain it briefly.

13

PHOTOSYNTHESIS

Microorganisms cannot solely depend on the oxidation of inorganic and organic compounds for energy. They may capture light energy and use it to synthesize ATP and NADH or NADPH. This process is termed as photosynthesis. In other words, photosynthesis can be defined as the trapping of light energy and its conversion to chemical energy, which is then used to reduce CO_2 and incorporate it in the organic form.

Photosynthesis is one of the significant metabolic processes on earth because almost all our energy is ultimately derived from solar energy. The synthesized ATP and NAD(P)H are necessary to manufacture organic materials of an organism. Over half of the photosynthesis is carried out by microorganisms. Photosynthesis involves two types of reactions. (i) light reactions and (ii) dark reactions. Light reaction involves the trapping of light energy and conversion to chemical energy and in dark reactions the chemical energy is then used to reduce or fix CO_2 and synthesize cell constituents.

LIGHT REACTION IN CYANOBACTERIA (PHOTOSYNTHETIC APPARATUS)

Role of Chlorophyll and Bacterial Chlorophyll in Photosynthesis

Chlorophyll structurally has a porphyrin ring as in cytochromes. Chlorophyll contains magnesium ion instead of iron at the centre of the porphyrin ring. There are two types of chlorophyll, chlorophyll *a* and chlorophyll *b*. The structure of chlorophyll *a* (Figure 13.1), the principal chlorophyll of higher plants, most algae, and the cyanobacteria is green in colour because it absorbs red and blue light and transmits green light.

Chlorophyll *a* shows strong absorption of red light at a wavelength of 680 nm and blue light at 430 nm.

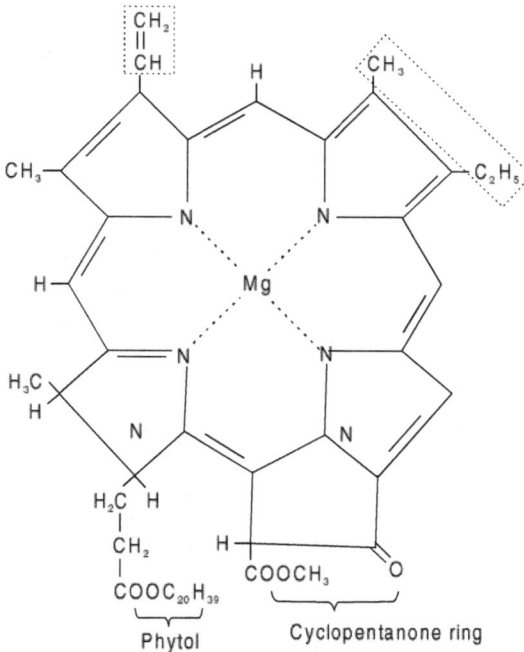

Figure 13.1   Structure of chlorophyll a.

There are different chlorophylls based on their chemical nature and that are distinguished by their different absorption spectra. Chlorophyll *b* for instance, absorbs maximally at 660 nm rather than 680 nm. Among prokaryotes, the cyanobacteria have chlorophyll. But the purple-green bacteria have a chlorophyll of slightly different structure, called bacteriochlorophyll.

Bacterial chlorophyll from purple phototrophic bacteria absorbs maximally at 720–790 nm and one type of bacterial chlorophyll absorbs at 1020 nm.

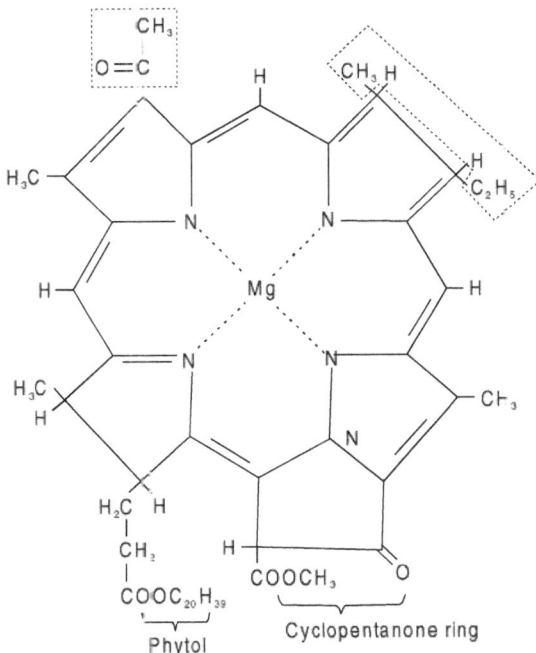

Figure 13.2 Structure of bacteriochlorophyll a.

Photosynthetic Membrane System

In prokaryotes, chloroplast (special intracellular organelles where photosynthesis takes place) are not present and photosynthetic pigments are integrated into internal membrane systems that arise from the following.

i. Invagination of the cytoplasmic membrane, e.g. Purple bacteria

ii. Cytoplasmic membrane itself, e.g. Heliobacteria

iii. Both the cytoplasmic membrane and specialized non-unit-membrane-enclosed structures called chlorosomes, e.g. Green bacteria.

In chlorosomes, antenna (chlorophyll-light harvesting) bacterial chlorophyll molecules are not bound to proteins as in purple bacteria but still they function to absorb light energy and transfer it to the reaction centre. Bacteriochlorophyll is present in the cytoplasmic membrane and other photosynthetic pigment are present in cyanobacteria. The most widespread pigment is carotenoid which is a long molecule usually yellowish in colour, that possesses an extensive conjugated double bond system. Red algae and cyanobacteria have photosynthetic pigments called phycobiliproteins, consisting of a protein with a tetrapyrrole attached.

Phycoerythrin is a red pigment with a maximum absorption around 550 nm. Phycocyanin is blue (maximum absorption at 620–640 nm).

Accessory Pigments

Chlorophyll cannot effectively absorb light in the blue-green through yellow range (470–630 nm). The accessory pigments absorb light in this range and transfer the energy to chlorophyll. Accessory pigments also protect microorganisms from intense sunlight which could oxidize and damage the photosynthetic apparatus in the absence of these pigments.

Antennas

Chlorophyll and accessory pigments are assembled in highly organized arrays called 'antennas'. The antennas usually consist of 300 chlorophylls which aid in capturing more light energy. This captured energy is transferred from chlorophyll to chlorophyll until it reaches the 'reaction centre chlorophyll.'

The reaction centre chlorophyll is directly involved in the photosynthetic electron transport chain. Based on oxidation, photosynthesis can be of two types:

Anoxygenic photosynthesis, e.g. Purple, green sulphur bacteria

Oxygenic photosynthesis, e.g. Cyanobacteria and higher plants.

In anoxygenic type, organisms obtain electron to reduce $NADP^+$ from H_2S and organic compounds. In oxygenic photosynthesis, they obtain electron for NADP reduction from the oxidation of water molecules and O_2 is released as a by-product. In eukaryotes and cyanobacteria, antennas are associated with two different photosystems.

Photosystem I

It absorbs longer wavelength (≥680 nm) and funnels the energy to a special chlorophyll *a* molecule called P700 (the term P700 signifies that this molecule most effectively absorbs light at a wave length of 700 nm).

Photosystem II

This system traps light of shorter wavelength (≤680 nm) and transfers its energy to the special chlorophyll P680.

When the photosystem I antenna transfers light to the reaction centre P700 chlorophyll, P700 absorbs energy and is excited. The reduction potential at this time becomes negative. Then it donates its excited or high-energy electron to a specific acceptor probably chlorophyll *a* molecule (A) or an iron–sulphur protein. The electron is even transferred to ferredoxin and can then travel in either of the two directions. In the cyclic pathway, the electrons move in a cyclic route through a series of electron carriers and back to oxidized P700*.

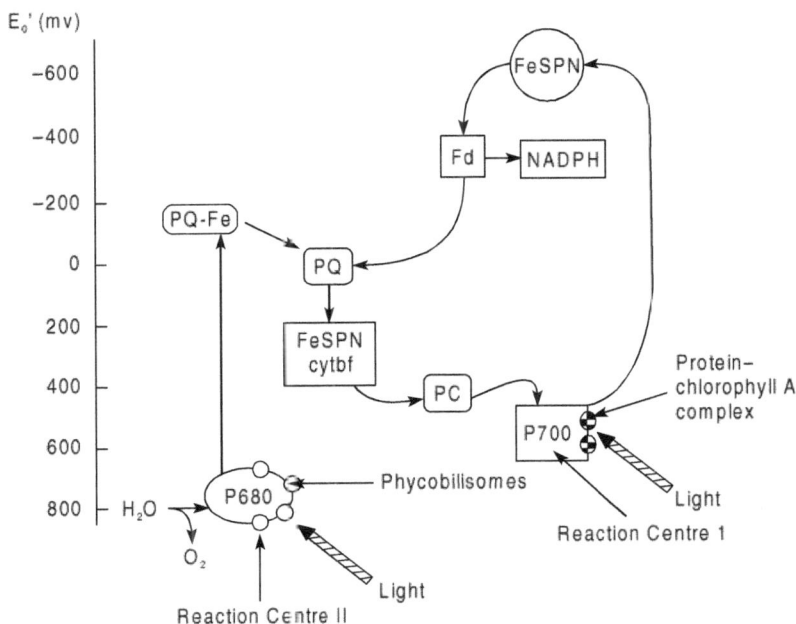

Figure 13.3   Electron transport in cyanobacteria.

The pathway way is termed as cyclic because the electrons from P700 return to P700* after travelling through the photosynthetic electron transport chain. ATP is formed during cyclic electron transport in the region of cytochrome b6. This is called cyclic phosphorylation because electrons travel in a cyclic pathway and ATP is formed.

The organisms possess two reaction centres that allow the trapping of radiant energy over a broad-energy spectrum. Phycobilisomes, which contain phycocyanin, phycoerythrin, allophycocyanin, and tetrapyrroles are the light-harvesting elements of reaction centre II, while protein-chlorophyll a complexes serve the same function in reaction centre I. Light energy trapped by either photosystem is used to eject an electron that reduces a low-oxidation potential carrier molecule, an iron–sulphur protein (FeSPN) for reaction centre I, and a plastoquinone-iron complex (PQ Fe) for reaction centre II. Electrons from FeSPN are transferred sequentially to ferredoxin (Fd), plastoquinone (PQ), an iron–sulphur–cytochrome b–cytochrome c complex, plastocyanin (PC), and, finally, back to reaction centre I. Electrons from PQ Fe are fed into PQ and allow net formation of NADPH with Fd as electron donor, providing reducing power for carbon dioxide reduction.

NONCYCLIC PHOTOPHOSPHORYLATION

The electrons can travel through noncyclic photophosphorylation. Photosystem I and photosystem II are involved in noncyclic photophosphorylation. In this pathway the P700 is excited and donates electron to ferredoxin as mentioned before. The reduced ferredoxin reduces NADP to NADPH. Because the electrons contributed to $NADP^+$ cannot be used to reduce oxidized P700 in this process, photosystem II participation is also required. It donates electrons to oxidized P700 and generates ATP in the process. The photosystem II antenna absorbs light energy and excites P680, which then reduces pheophytin *a*. Pheophytin *a* is chlorophyll *a* in which two hydrogen atoms replace the central magnesium. Electrons subsequently travel to Q (probably a plastoquinone) and down the electron transport chain of P700. The oxidized P680 then obtains an electron from the oxidation of H_2O to O_2. Thus, electron flows from water all the way to $NADP^+$ with the aid of energy from photosystems, and ATP is synthesized by noncyclic photophosphorylation. It appears that one ATP and one NADPH are formed when two electrons travel through the noncyclic pathway.

The dark reactions require 3 ATP and 2 NADPH to reduce 1 CO_2 and use it to synthesize carbohydrates.

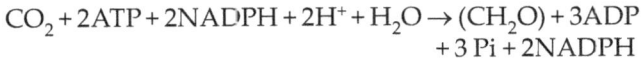

$$CO_2 + 2ATP + 2NADPH + 2H^+ + H_2O \rightarrow (CH_2O) + 3ADP + 3 Pi + 2NADPH$$

The noncyclic system generates one NADPH and one ATP per pair of electrons; therefore 4 electrons passing through the system will produce 2 NADPHs and 2 ATPs. A total of 8 quanta of light energy (4 quanta of each photosystem) is needed to propel the 4 electrons from water to $NADP^+$. Since the ratio of ATP to NADPH required for CO_2 fixation is 3 : 2, at least one more ATP molecule must be supplied. Cyclic photophosphorylation probably operates independently to generate the extra ATP. This requires absorption of another 2 to 3 quanta of energy. It follows that around 10–12 quanta of light energy are needed to reduce and incorporate one molecule of CO_2 during photosynthesis.

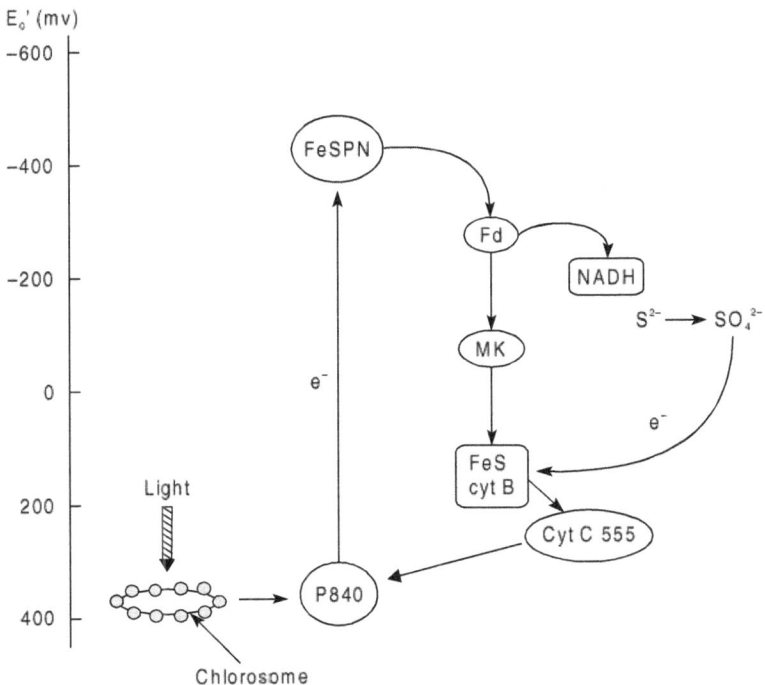

Figure 13.4 Electron transport in the green bacteria.

Light energy is trapped in a chlorosome vesicle adjacent to the cytoplasmic membrane. The trapped energy is transferred to a reaction centre, P840, in the cytoplasmic membrane. Electrons ejected from P840 are transferred sequentially to an iron-sulphur protein (FeSPN), ferredoxin (Fd), menaquinone (MK), an iron sulphur-cytochrome b complex (FeS cyt b), a cytochrome c (cyt c), and are finally returned to P840. Sulphide oxidation to sulphate, with donation of electrons to FeS-cyt b, provides reducing power for carbon dioxide reduction, with Fd serving as the immediate donor of electrons to NAD.

LIGHT REACTION IN GREEN AND PURPLE BACTERIA

Green and purple photosynthetic bacteria differ from cyanobacteria in that they do not use water as an electron source to produce O_2 photosynthetically, i.e., they are anoxygenic (photosynthesis that does not use water to produce O_2). NADH is not directly produced in the photosynthetic light reaction of purple bacteria.

Green bacteria can reduce NAD^+ directly during the light reaction to synthesize NADH and NADPH. Green and purple bacteria must use electron donors like hydrogen, hydrogen sulphide, elemental sulphur, and organic compounds which have more negative reduction potentials than water and are therefore easier to oxidize (better electron donors). Finally green and purple bacteria possess slightly different photosynthetic pigments like bacterial chlorophyll. Bacterial chlorophyll a and b have maximum at 775 and 790 nm respectively. In vivo maxima are about 830 to 890 nm (bacterial chlorophyll a) and 1020 to 1040 nm (bacterial chlorophyll b). Many differences found in green and purple bacteria are due to their lack of photosystem II. These bacteria cannot use water as an electron donor in noncyclic electron transport. Without this photosystem II they cannot produce O_2 from H_2O photosynthetically and are restricted to cyclic photophosphorylation. Indeed, almost all purple and green sulphur bacteria are strict anaerobes. When the special reaction centre chlorophyll 870 is excited, it donates an electron to bacteriopheophytin. Electrons then flow to quinones and through an electron transport chain back to P870 while driving ATP synthesis. Green and purple bacteria face a further problem because they also require NADH or NADPH for CO_2 incorporation. They may synthesize NADH in at least three ways.

1. If they are growing in the presence of hydrogen gas, which has a reduction potential more negative than that of NAD^+, the hydrogen can be directly used to produce NADH.

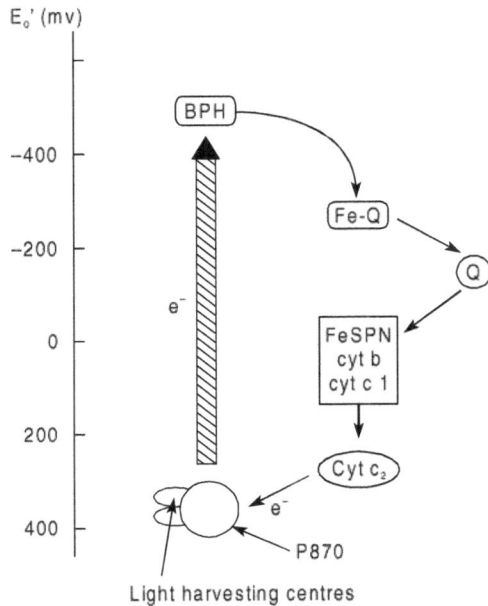

Figure 13.5 The reactions of electron transport in purple bacteria.

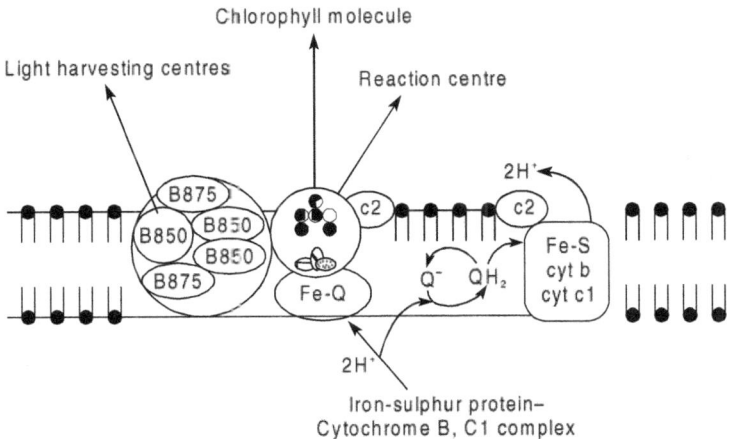

Figure 13.6 Structural relationships among the electron transport components of the purple photosynthetic bacteria.

2. Like chemolithotrophs, many photosynthetic purple bacteria use ATP or proton motive force to reverse the flow of electrons

in an electron transport chain and move them from inorganic to organic donors to NAD⁺.

3. Green and sulphur bacteria appear to carry out a simple form of noncyclic photosynthetic electron flow to reduce NAD⁺.

STUDY OUTLINE

- Microorganisms derive energy only from the oxidation of inorganic and organic components. They may capture light energy and use it to synthesize ATP and NADH or NADPH. This process is termed as photosynthesis.

- Photosynthesis can be divided into two parts (i) Light reactions and (ii) Dark reactions.

- Light reaction involves the trapping of light energy and conversion to chemical energy and in the case of dark reaction, the chemical energy is then used to reduce or fix CO_2 and synthesize cell constituents.

- For photosynthesis, the pigment chlorophyll is used. In the case of bacteria, two types chlorophyll are present—chlorophyll *a* and chlorophyll *b*.

- In photosynthesis, generally photosystem I and photosystem II are involved.

- Based on their oxidation, photosynthesis includes anoxygenic photosynthesis and oxygenic photosynthesis.

CONCEPT CHECK

1. Write about the pigments involved in photosynthesis.

2. What are oxygenic and anoxygenic photosynthesis?

3. Briefly explain about photosystems I and II.

CRITICAL THINKING

How can you differentiate photosynthesis of bacteria from photosynthesis of plants?

14

CARBON DIOXIDE FIXATION

INTRODUCTION

Fixation of carbon dioxide may be defined as the process by which inorganic carbon dioxide becomes incorporated (fixed) into the structure of organic compounds within the cells. This is the basis of autotrophic metobolism.

CALVIN CYCLE

CO_2 fixation occurs within many autotrophic microbial and plant cells via a metabolic pathway known as calvin cycle. Calvin cycle involves the synthesis of glyceraldehyde 3-phosphate from CO_2. It effectively takes three turns of the calvin cycle with one CO_2 molecule entering at each turn to synthesize one molecule of glyceraldehyde 3-phosphate, because glyceraldehyde 3-phosphate contains three carbon atoms. Thus the Calvin cycle is sometimes referred to as C_3 pathway.

CO_2 is the most oxidized form of carbon, and its conversion to glyceraldehyde 3-phosphate via the Calvin cycle requires a great deal of energy (ATP) and reducing power (NADPH). The overall equation for this process is

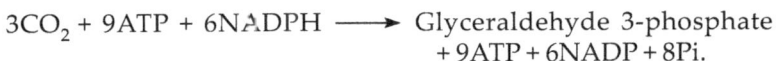

$$3CO_2 + 9ATP + 6NADPH \longrightarrow \text{Glyceraldehyde 3-phosphate} + 9ATP + 6NADP + 8Pi.$$

The Calvin, or carbon reduction cycle is the main metabolic pathway used by autotrophs for the conversion of carbon dioxide to organic carbohydrates. The pathway, which is active in photoautotrophs and chemolithotrophs, requires the input of carbon dioxide, ATP (energy), and NADPH (reducing power).

In photoautotrophs, the ATP and NADPH come from the light reactions of photosynthesis. In chemoautotrophs, (chemolithotrophs) the ATP and NADPH are derived from the oxidation of inorganic compounds. The Calvin cycle is known as a dark reaction because, although it requires ATP and NADPH and can be a part of the overall process in photosynthesis, it does not involve any reaction directly coupled with the input of light energy. Calvin cycle can proceed in the absence of light if adequate ATP and NADPH are supplied.

The initial metabolic step in the Calvin cycle involves the reaction of CO_2 with ribulose 1,5-bisphosphate to form an unstable 6-carbon compound that immediately splits to form two molecules of 3-phosphoglycerate. This reaction is highly exergonic with a ΔG^0 of –12.4 kcal/mole. The reaction of CO_2 with ribulose 1,5-bisphosphate is catalysed by ribulose 1,5-bisphosphate carboxylase (RuBisCo), a key enzyme in the calvin cycle. Many autotrophic bacteria store ribulose 1,5-bisphosphate carboxylase as inclusions within the cell. These inclusion bodies called carboxysomes, are polyhedral structures that contain insoluble, crystalline ribulose 1,5-bisphosphate carboxylase. Carboxysomes are found in nitrifying bacteria, the photosynthetic cyanobacteria and in many of the sulphur-oxidizing autotrophic bacteria.

When the reaction catalysed by ribulose 1,5-bisphosphate carboxylase occurs three times, it allows three molecules of CO_2 to react with three molecules of ribulose 1,5-bisphosphate and generates a total of six molecules of 3-phosphoglycerate. Five of the six molecules of 3-phosphoglycerate go through a series of reactions that regerarate the original three molecules of ribulose 1,5-bisphosphate. The one remaining 3-phosphoglycerate molecule is reduced to form the net product of the cycle, the glyceraldehyde 3-phosphate molecule. It is because the ribulose 1,5-bisphosphate is regenerated in this pathway that this process is called a cycle. The only net carbon flow is the entry of three CO_2 molecules, accompanied by the production of one molecule of glyceraldehyde 3-phosphate. Carbon dioxide continually flows in to the calvin cycle and glyceraldehyde 3-phosphate continually flows out, all mediated by the interconversion of the cycle's intermediates.

The glyceraldehyde 3-phosphate molecules that are formed during the Calvin cycle can further react to form glucose and polysaccharides composed of linked glucose units such as starch and cellulose. It takes six turns of the Calvin cycle to form one 6-carbon carbohydrate such as glucose. The overall conversion of CO_2 to glucose is highly endergonic and requires 114 kcal/mole. For this conversion, the input of energy in the form of ATP is 18 molecules and the reducing power via NADPH is 12 molecules. To satisfy the ATP and NADPH requirements of this process in algae and cyanobacteria, light photoacts are needed (reaction is which a photon is absorbed), four each in photosystems I and II. Since one molecule of photon is approximately equivalent to 47 kcal, the efficiency of photosynthesis is about 114 (47 × 8) or 30%.

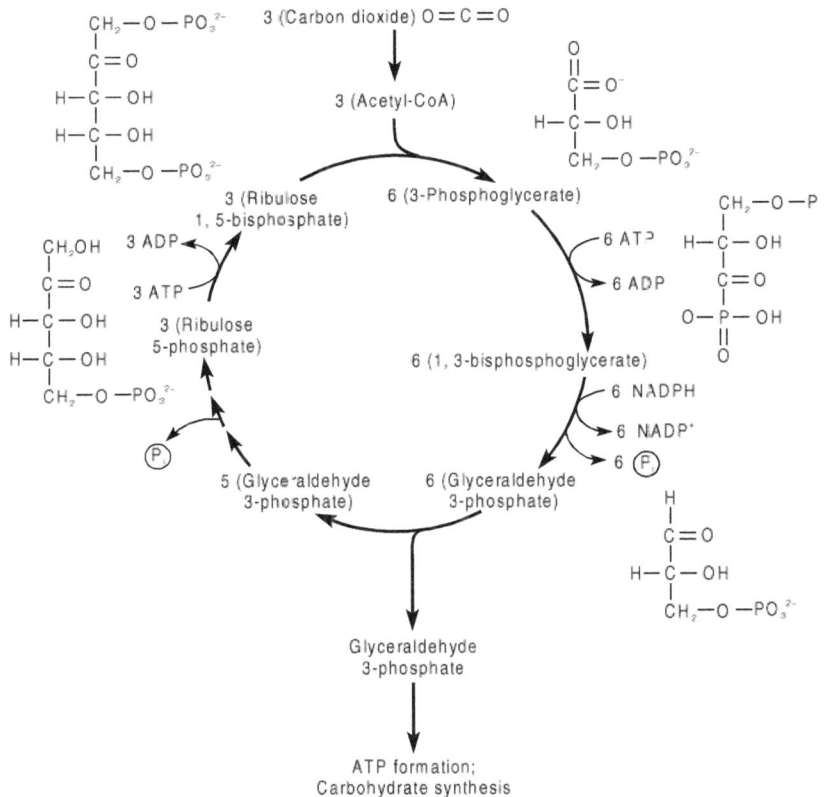

Figure 14.1   Calvin cycle.

REDUCTIVE CARBOXYLIC ACID CYCLE OR REVERSE TCA CYCLE

Some phototrophic organisms such as green sulphur bacterium *Chlorobium*, fix CO_2 via a reverse (reductive) tricarboxylic acid cycle.

This cycle starts with the reduced conversion of oxaloacetate to malate which is then converted into fumarate. The fumarate is further reduced to succinate. With the aid of ATP, succinate is then converted into succinyl-CoA. In this step CO_2 is then added with succinyl CoA by a reduced ferredoxin-linked enzyme, which forms α-ketoglutarate. Then a second molecule of CO_2 is reductively added to the α-ketoglurarate to form isocitric acid (isocitrate) and then citric acid (citrate). The formed citrate then splits into oxalacetate and acetyl-CoA with the input energy of ATP. The formed oxaloacetate, then again enters into the next cycle. The acetyl-CoA formed by this reverse TCA cycle, containing the two fixed carbon atoms can be reductively carboxylated by another reduced ferredoxin-linked enzyme to form a pyruvate. With the help of ATP, and in the final step, another CO_2 is fixed to convert the phosphoenol pyruvate into oxalocetate. In collection, four steps of CO_2 molecules are needed for each reverse TCA cycle. And thus 4 CO_2 molecules are fixed in this cycle. This is carried out with the help of 3 ATP molecules. The enzymes involved in reverse TCA cycle are

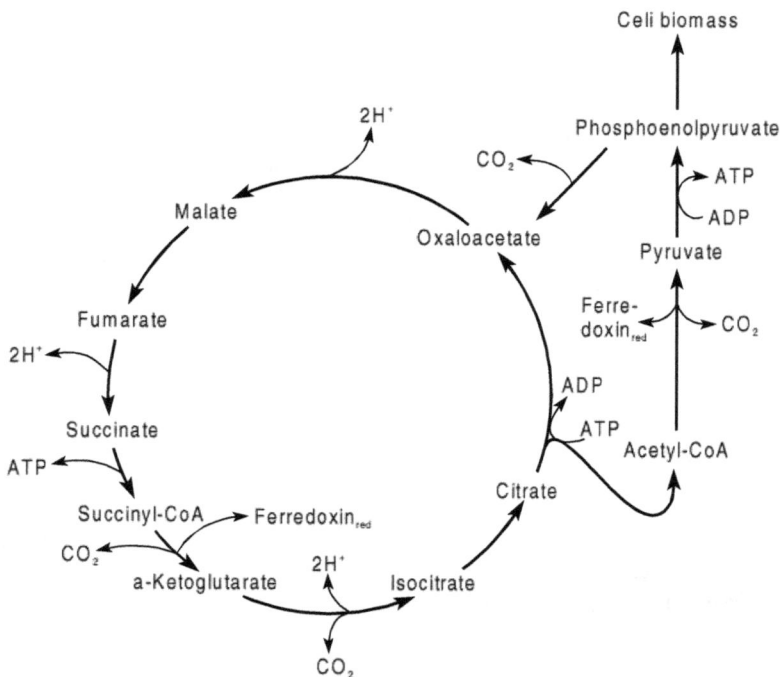

Figure 14.2   Reductive tricarboxylic acid cycle.

same as those involved in the normal TCA cycle with one exception—the enzyme citrate lyase which cleaves citrate into acetyl-CoA and oxaloacerate direction, citrate is produced from acetyl-CoA and oxaloacetate by the enzyme citrate synthetase.

The reductive tricarboxylic acid cycle of some photoautotrophs converts CO_2 molecules into an oxaloacetate molecule for incorporation in cell biomass.

C4 PATHWAY IN MICROORGANISMS

A common pathway for fixation of CO_2 in heterotrophs and autotrophs is the C4 pathway. It is so designated beause the product formed via this pathway, oxaloacetate, is a 4-carbon molecule. In this metabolic pathway, pyruvate or phosphoenol pyruvate (metabolites of the glycolytic pathway) react with CO_2 to form oxaloacetate, an

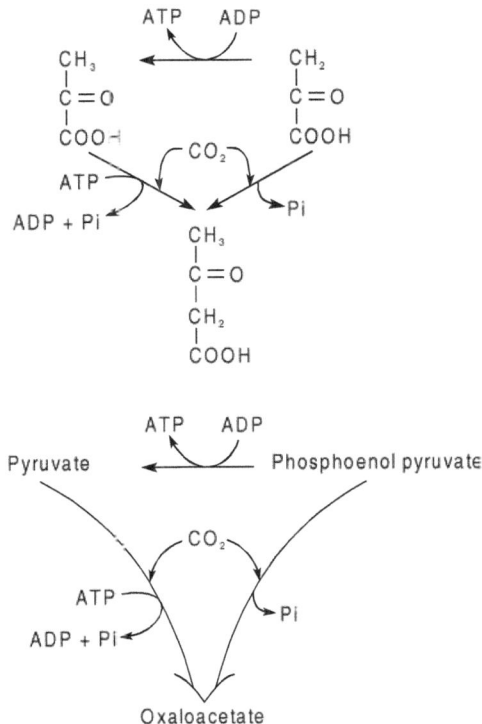

Figure 14.3   Hydroxypropionate pathway for CO_2 fixation.

intermediate metabolite of the TCA cycle. The oxaloacetate formed in this pathway can then be used in amino acid and nucleic acid biosyntheses. Although all microorganisms fix CO_2 as a part of their metabolism, heterotrophic organisms are unable to form a significant portion of their macromolecules from the C4 pathway alone, and so remain dependent on organic compounds which form their substrate for cellular growth.

C_4 pathway adds CO_2 to either pyruvate or phosphoenol pyruvate to produce oxaloacetate.

The green nonsulphur bacterium for example *Chloroflexus* grows autotrophically using H_2 or H_2S as an electron donor but it does not use Calvin cycle or reverse TCA cycle to fix CO_2 into inorganic carbon. Instead, two molecules of CO_2 are fixed and it is converted into one

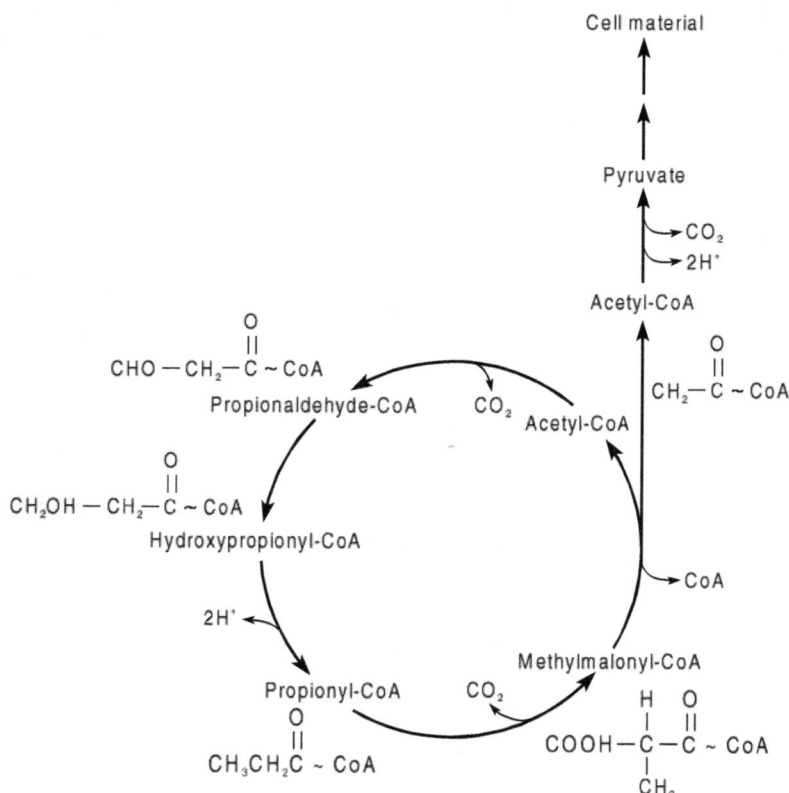

Figure 14.4   Hydroxypropionate pathway.

molecule of acetyl-CoA through the hydroxypropionate pathway. In this pathway the hydroxypropionyl-CoA is a specific intermediate. The acetyl-CoA formed by this pathway can then be reduced and carboxylated to form pyruvate. The net result is that three CO_2 molecules are converted into one pyruvic acid molecule.

The hydroxypropionate pathway of *Chloroflexus* converts CO_2 into acetyl-CoA for incorporation into cell biomass.

STUDY OUTLINE

- Fixation of carbon dioxide may be defined as the process by which inorganic carbon dioxide becomes incorporated (fixed) into the structure of organic compounds within the cells.

- Carbon dioxide fixation occurs through calvin cycle, reductive or reverse TCA cycle, C_4 cycle, hydroxypropionate cycle.

- Calvin cycle is otherwise known as a dark reaction because, although it requires ATP and NADPH and can be a part of the overall process in photosynthesis, it does not involve any reaction directly coupled with the input of light energy.

- Some green sulphur bacteria like *Chlorobium*, fix CO_2 via a reverse (reductive) tricarboxylic acid cycle. This cycle starts with the reduced conversion of oxaloacetate to malate.

- C_4 pathway is a common pathway for the fixation of CO_2 in heterotrophs and autotrophs.

CONCEPT CHECK

1. Write about Calvin cycle in detail.
2. How do microorganisms fix CO_2 from various pathways?
3. Expalin the C_4 cycle in microorganisms.

15

OXIDATIVE PHOSPHORYLATION AND ELECTRON TRANSPORT CHAIN

Throughout the microbial world, a number of mechanisms exist for substrate-level phosphorylation, but electron transport is an equally prominent mechanism for ATP formation. Electron transport was originally thought to be restricted to oxybiontic organisms, that is, organisms that use molecular oxygen in their metabolism. Thus, initially, organisms were regarded as being of one of the two types, respiratory or fermentative. Respiratory organisms were regarded as those organisms that generated energy obligatorily, by electron transport, with oxygen as the terminal electron acceptor. Fermentative organisms, by contrast, were those organisms that could not generate energy by electron transport, could not use molecular oxygen in energy metabolism, and were thus restricted to substrate-level mechanisms for energy generation.

The electron transport systems are composed of membrane-bound electron carriers. These systems have two basic functions (i) to accept electron from an electron donor and transfer them to an electron acceptor (ii) to conserve some of the energy released during electron transfer for synthesis of ATP.

There are several types of oxidation and reduction enzymes. Some of them are listed below.

NADH DEHYDROGENASE

This enzyme transfers hydrogen atoms from NADH. The enzymes are protein-bound to the inner surface of the cell membrane. They accept hydrogen atoms from NADH, generated during various cellular reactions, and pass the hydrogen atoms to flavoproteins.

NAD   **NADH + H***

$2H\ (2H^+ + 2e^-)$
Reduced
Oxidized

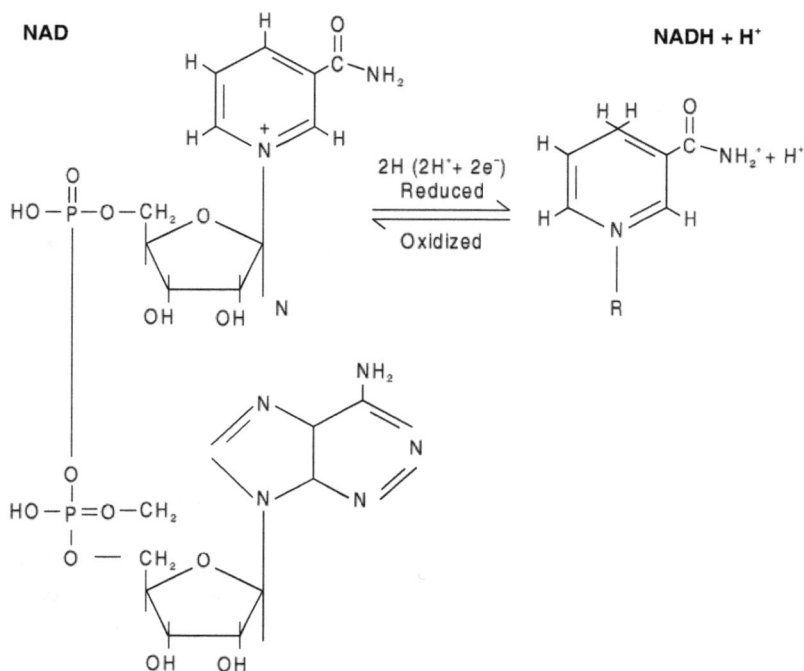

FLAVOPROTEINS

Riboflavin contains electron carriers, generally called flavoprotein (which contain flavin mononucleotide (FMN) or flavin adenine dinucleotide). Flavoproteins are proteins containing a derivative of riboflavin, the flavin protein, which is bound to a protein in the prosthetic group, is alternatively reduced when it accepts hydrogen atoms and is oxidized when electrons are passed on.

$$2H \longrightarrow \boxed{2H^+} + 2e^-$$

Note that protein accepts hydrogen atoms and donates electrons. Now, what happened to the protons? We will see that later.

Flavoproteins generally exist in two forms.

FMN – Flavin mononucleotide is bounded to ribose sugars and adenine through a second phosphate.

FAD – Flavine adenine dinucleotide is bounded to two ribose sugars and adenine through a second phosphate.

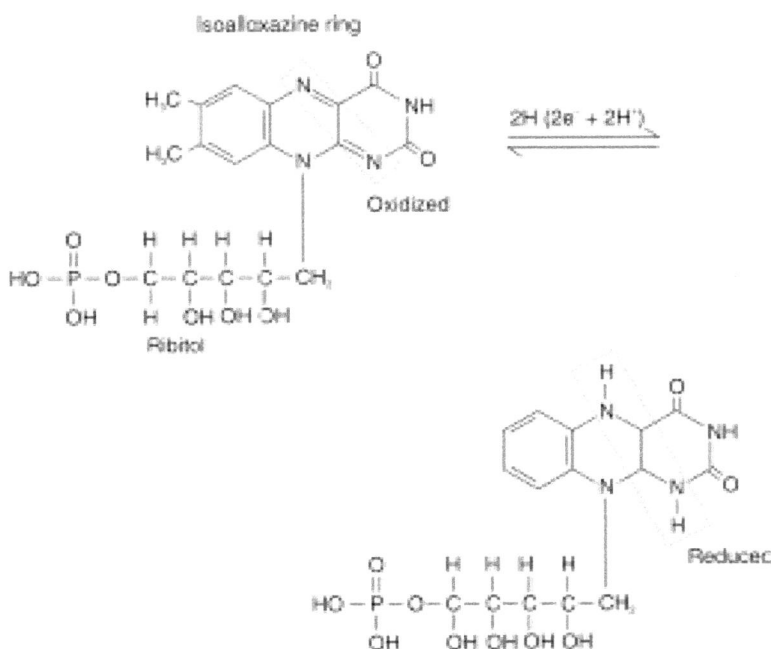

Figure 15.1   Structure of flavin mononucleotide.

Riboflavin, also called vitamin B_2 is a required growth factor for some organisms.

CYTOCHROMES OR CYT a, b, c

Cytochromes are proteins containing a non-porphyrin ring called heme (prosthetic group). They undergo oxidation–reduction through loss or gain of a single electron by the iron atom at the centre of the cytochrome.

$$\text{Cytochrome–Fe}^{2+} \rightleftharpoons \text{cytochrome-Fe}^{3+} + e^-$$

There are various subclasses of cytochromes like cytochromes a, b, c depending on their positive and negative reduction potential. Sometimes cytochromes bind tightly to one another or with iron–sulphur proteins, e.g. cytochrome bc complex, which contains two different types—b-type cytochromes and c-type cytochromes. This cytochrome plays an important role in energy metabolism.

NON-HEME IRON–SULPHUR PROTEINS

Proteins are associated with the electron transport chain in various places, e.g. Ferredoxin, a common iron–sulphur protein in biological systems, has an Fe_2S_2 configuration or Fe_4S_4. The iron atoms are bounded to the protein via sulphur atoms from cysteine residues. Like cytochromes, iron–sulphur proteins carry electrons only and not hydrogen atoms. The reduction potentials of iron–sulphur proteins vary over a wide range depending on the number of iron and sulphur atoms.

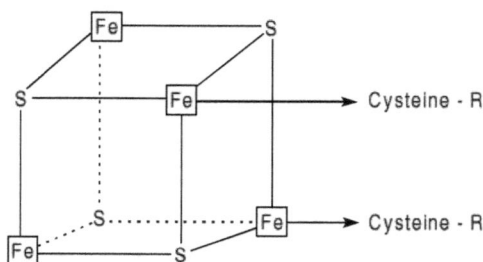

Figure 15.2 Structure of Fe_4S_4.

QUINONES

Quinone is a nonprotein electron carrier and liquid-soluble compound, sometimes called coenzyme Q. Its molecules are highly hydrophobic involved in electron transport. Like flavoproteins, quinones serve as hydrogen atom acceptor and electron donor.

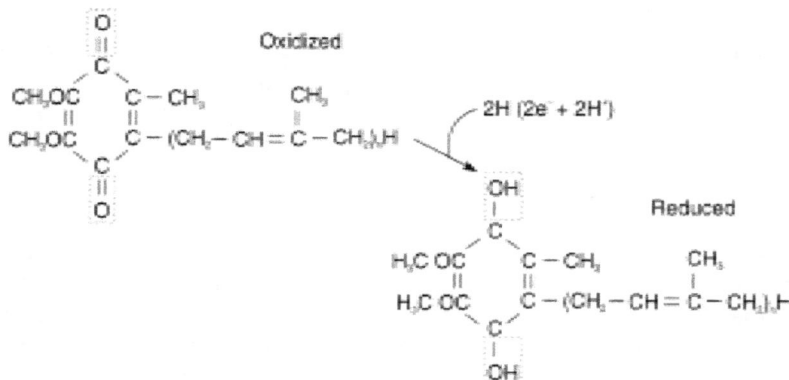

ENERGY CONSERVATION FROM ELECTRON TRANSPORT

During electron transport, ATP is produced by the process of oxidative phosphorylation. The production of ATP is linked directly to the establishment of *proton motive force* across the membrane, electron transport reactions, serving to establish their energized state to the membrane.

THE PROTON MOTIVE FORCE—CHEMIOSMOSIS

While discussing about the structure of electron transport chain, it should be clearly understood that all the proteins which are needed for the electron transport chain are present in the cell membrane. The protein can access both the outside and the inside of the cell. They are called as transmembrane proteins. The electron transport carriers are oriented in such a way that a separation of protons from electrons occurs during the transport process. When the hydrogen atom enters the electron transport chain ($2H \rightarrow 2e^- + 2H^+$), electrons enter the electron transport carriers and the protons are accumulated outside the cell (environment). In gram-negative bacteria they are accumulated in the periplasm. This will create an external acidic environment. The electrons which passed through the electron transport chain reach the terminal acceptor (in aerobic organisms, the acceptor is O_2). During reduction of O_2, one water molecule is formed. This is done by splitting the available water molecule to H^+ and OH^-. The H^+ in the reaction combines with O_2 giving H_2O. The OH^- in this reaction accumulates inside. So the inside OH^- and the outside H^+ cannot pass freely in and out of the cell membrane. So this cannot be used to maintain the neutrality of the cells. This creates a pH gradient and electrochemical potential (i.e., H^+ has positive charge and is acidic; whereas OH^- has negative charge and is basic). This electrochemical potential energizes the cell membrane and this energy is used by the cell. The ATP during oxido-reduction is synthesized because each reaction releases some energy and is utilized for ATP production. ATP is also synthesized by proton motive force. The proton gradient driving ATP synthesis was first proposed as the chemiosmotic theory in 1961 by the English scientist Peter Mitchell. Later he received the Nobel Prize for his important contribution.

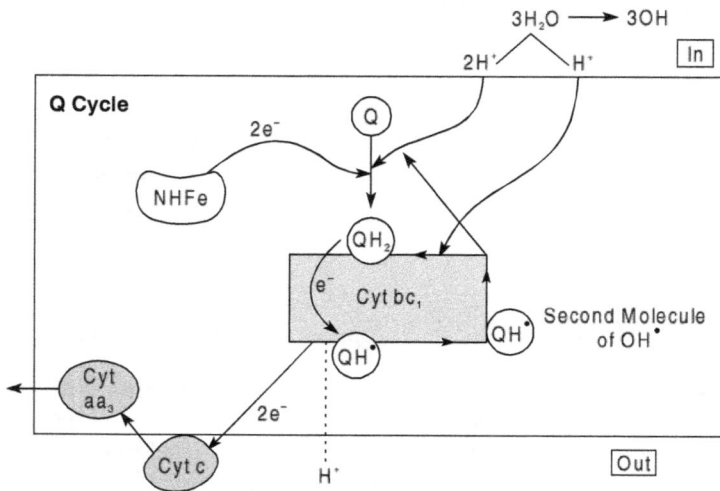

Figure 15.3   Q cycle in electron transport chain.

GENERATION OF PROTON MOTIVE FORCE

In electron transport chain NADH donates two electrons to the NADH dehydrogenase (NDH). In the second step this $2e^-$ and $2H^+$ enter into the next enzyme FADH–dehydrogenase (FDH) complex. Both enzymes can accept both electrons and protons, and thus activate the third enzyme. The third enzyme is a non-heme iron–sulphur protein (NHFe). This protein cannot carry protons. It has the ability to carry only electrons. This NHFe protein donates electrons to coenzyme Q (QH_2/ Q). Simultaneously the coenzyme accepts the protons by the dissociation of water molecule. Then the coenzyme Q donates one electron to the cytochrome bc_1 complex. Cytochrome bc_1 donates electrons to cytochrome linked to the translocation of protons across the membrane. The cytochrome bc_1 complex is oriented in the membrane in such a way that protons are discharged to the environment when electrons are transferred to an acceptor, resulting in the accumulation of OH^- in the cytoplasm and protons on the outer surface of the membrane.

What is the role of Q cycle in ETC and how does it take place?

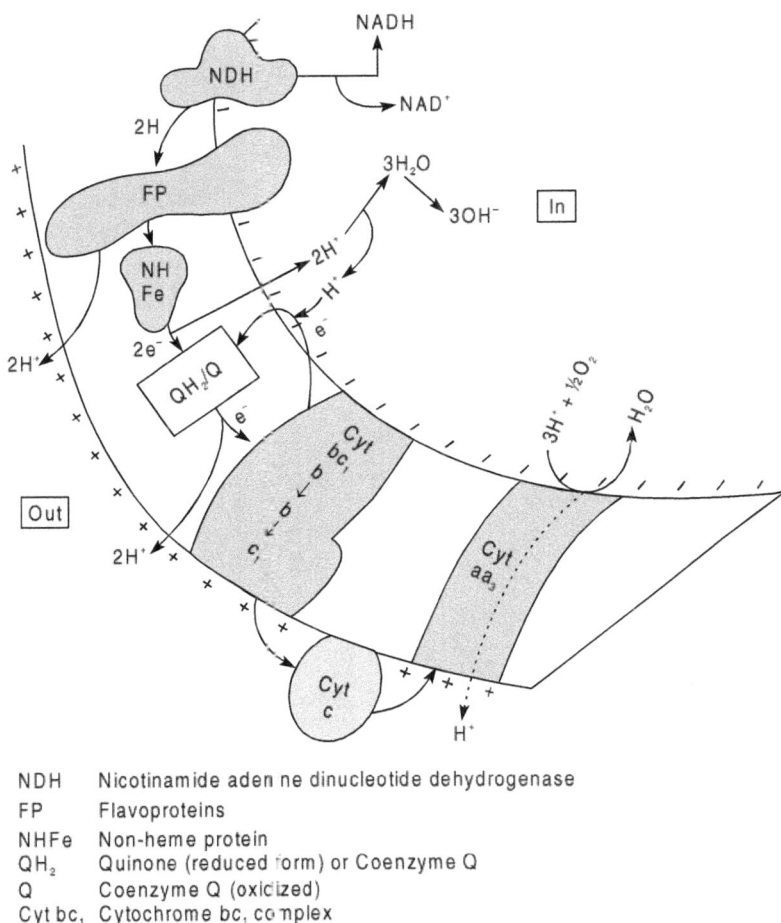

NDH   Nicotinamide adenine dinucleotide dehydrogenase
FP   Flavoproteins
NHFe   Non-heme protein
QH_2 Quinone (reduced form) or Coenzyme Q
Q   Coenzyme Q (oxidized)
Cyt bc_1 Cytochrome bc, complex

Figure 15.4   Generation of proton motive force.

In the Q cycle, the quinone is reduced to give QH_2. This QH_2 donates one electron to cytochrome bc_1 and one proton across the membrane. After this process, the QH_2 becomes $QH^\bullet$. This $QH^\bullet$ then accepts one electron from cytochrome bc_1 and the QH is reduced, forming QH_2. If two QH enter into the cycle, one QH is reduced as QH_2 and the other is oxidized to Q. This is called as Q cycle which acts to increase the number of protons extended across the membrane at the Q-bc_1 site. Electrons travel from the bc_1 complex to cytochrome c and cytochrome a, and the latter serves in conjunction with the terminal oxidase. Finally, the reduction of $\frac{1}{2} O_2 + H_2O$ occurs and the electron transport reactions are completed.

PROTON MOTIVE FORCE AND ATP FORMATION

An important component of this process is a membrane-based enzyme called ATP synthase, or ATPase (for short), which contains two main parts.

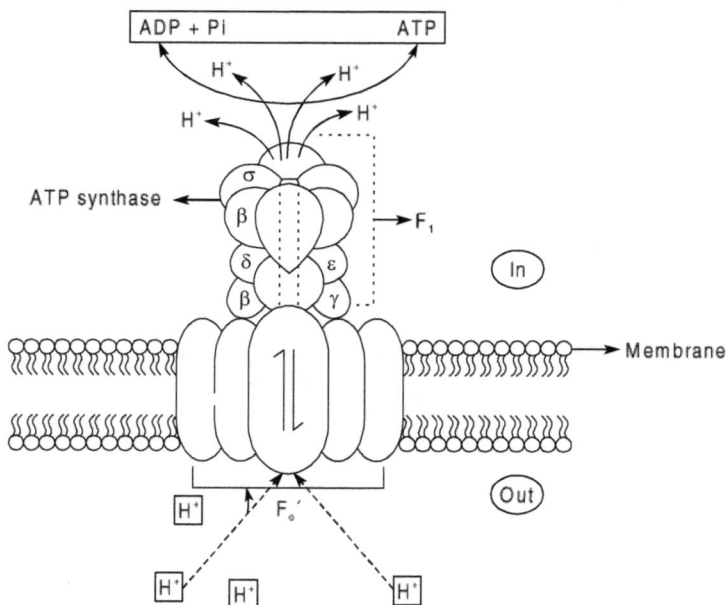

Figure 15.5   Structure of ATP synthase and proton motive force.

A multi subunit head piece present on the inside of the membrane and this enzyme catalyses a reversible reaction between ATP and ADP + Pi (inorganic phosphate). Operating in one direction, this enzyme catalyses the formation of ATP by controlled re-entry of protons across the energized membrane. The formation of proton gradient is energy-driven, the controlled dissipation of proton are energy-releasing, and some of the energy is conserved in the synthesis of ATP in a process called "oxidative phosphorylation". The mechanism of re-entry of proton to drive ATP synthesis across the cell membrane is not known. But the process is thought to involve a conformational change in ATPase proteins catalysed by protein translocation and the return of ATPase to its original conformation, releasing energy that is coupled to the synthesis of ATP. The β subunit of the ATPase is the catalytic site for the ATP synthesis. The β subunit is present in F_1 site of the ATPase. As protons enter from the outside, the β subunit rotates relative to $F_o{}'$ and this catalyses the reaction ADP + Pi $\rightarrow$ ATP. For the production of single ATP, it needs $4H^+$ protons re-entry across the membrane.

STUDY OUTLINE

• Respiratory organisms produce energy through the electron transport chain with oxygen as the terminal electron acceptor.

• Fermentative organisms, by contrast, are those organisms that cannot generate energy by electron transport, cannot use molecular oxygen in energy metabolism, and are thus restricted to substrate-level mechanisms for energy generation.

• Various enzymes like NADH dehydrogenases, flavoproteins, cytochromes, nonheme iron–sulphur proteins, and quinones are involved in electron transport chain.

• The formation of proton gradient is energy-driven, the controlled dissipation of proton is energy-releasing and some of the energy is conserved in oxidative phosphorylation.

• In ATP generation, Q cycle plays an important role in the transport of electrons.

CONCEPT CHECK

1. Define electron transport chain.

2. Explain about the various electron carriers and enzymes in the electron transport chain.

3. What is Q cycle? Explain the role of Q cycle in electron transport chain.

4. How is ATP generated through electron transport chain and explain about the proton motive force.

5. Outline the functions of ATPase enzyme in the generation of ATP.

CRITICAL THINKING

In some metabolic processes, ATP is directly synthesized without electron transport chain (ETC). Why then does a cell require (ETC)?

16

BIOLUMINESCENCE

Microbial luminescence is an interesting aspect of deep-sea life. Some bacteria produce light flashes when they are agitated by wave action or by a boat's wake. Many bacteria are also luminescent and have showed symbiotic relationship with fish dwelling in the benthic region. This relationship will be used by the fish. Sometimes fish use the glow of their resident bacteria as an aid in attracting and capturing prey in the complete darkness of the ocean depth. These organisms have an enzyme called 'luciferase'. Bioluminescence is the emission of light from the electron transport chain of certain living organisms. The enzyme luciferase picks up electrons from flavoproteins in the electron transport chain and emits some of the electronic energy as photons of light. Many fungi, marine microorganisms, jelly fish and crustaceans as well as the firefly are capable of generating bioluminescence, which requires considerable amount of energy. In the case of firefly, ATP is used in a set of reactions that convert chemical energy into light energy. The emission of light by luciferase enzyme involves the presence of luciferin, a complex carboxylic acid. The generation of light flash requires activation of luciferin by an enzymatic reaction with ATP in which a pyrophosphate cleavage of ATP occurs, to form a luciferyl adenylate.

This luciferyl adenylate is then acted upon by molecular oxygen and luciferase which bring about the oxidative decarboxylation of the luciferin to yield oxyluciferin. This reaction which has intermediate steps is accompanied by emission of light. The colour of the light flash differs with firefly species and appears to be determined by difference in the structure of the luciferase. Luciferin is then regenerated from oxyluciferin in the subsequent series of reactions. Other bioluminescent organisms use other types of enzymatic reactions to generate light.

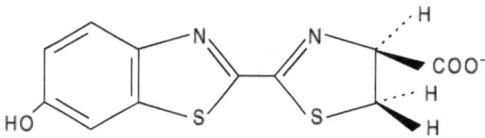

Structure of firefly luciferin

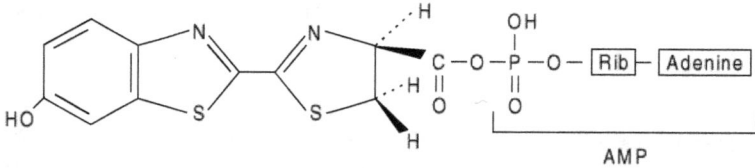

Structure of luciferyl adenylate

In the laboratory, pure firefly luciferin and luciferase are used to measure minute quantities of ATP by the intensity of the light flash produced. As little as a few picomoles (10^{-12} mol) of ATP can be measured in this way.

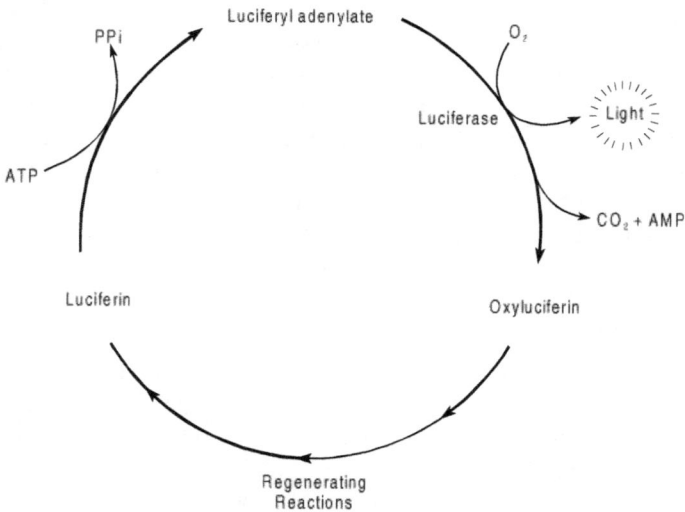

Figure 16.1   Important components in firefly bioluminescence, and the firefly bioluminescence cycle.

INDUSTRIAL APPLICATION OF BIOLUMINESCENCE

Bioluminescent bacteria can act as biosensors—bacteria that can locate biologically active pollutants. Biosensors do not require costly chemical or equipments, and they work quickly within minutes. To work, these bacterial biosensors require both a receptor that is activated in the presence of pollutant and a reporter that will make such a change apparent. Example for reporter is *Vibrio* or *Photobacterium*. Biosensors use the lux operon from these bacteria. This operon contains inducers and structural genes for the enzyme luciferase.

In the presence of a coenzyme called $FMNH_2$, luciferase reacts with the molecules in such a way that the enzyme–substrate complex emits blue-green light, which then oxidises the $FMNH_2$ to produce FMN. Therefore, a bacterium containing the lux gene will emit visible light when its receptor is activated. The lux operon is readily transferred to many bacteria. It has been transferred to the plant pathogen *Xanthomonas* in order to monitor the progression of its infection in plants. Because the altered bacteria emit light, their presence can be detected by photographic film.

STUDY OUTLINE

- Microbial luminescence is an interesting aspect of deep-sea life.

- Bioluminescence is the emission of light from the electron transport chain of certain organisms.

- The enzyme nciferase is mainly involved in this process.

CONCEPT CHECK

Give an outline note on bioluminescence.

17

PASTEUR EFFECT

Anaerobic respiration is much more effective than anaerobic process not involving electron transport and oxidative phosphorylation. Many microorganisms, when moved from anaerobic to aerobic condition, will drastically reduce their rate of sugar catabolism and switch over to aerobic respiration, a regulatory phenomenon known as the Pasteur effect.

Pasteur effect can be defined as the slower rate of glucose utilization by a microorganism which is growing aerobically by respiratory metabolism than by the same organism growing anaerobically which reflects the feedback inhibition; organisms which are capable of both fermentative and respiratory metabolism, inhibit glucose utilization on exposure to oxygen. For example, in a resting yeast cell suspension that is fermenting glucose, the introduction of oxygen results in the cessation of ethanol formation. This has been studied extensively in the attempt to elucidate the mechanism involved. Growing yeast cells do not show a noticeable Pasteur effect. Growing cells respire only 3 to 20% of the catabolized sugar and the rest is fermented. The mechanism involved in the cessation of fermentation in resting cell (i.e. under conditions of nitrogen starvation) is a progressive inactivation of the sugar transport system. As a consequence, the contribution of respiration of glucose catabolism which is small in growing cells, becomes quite significant under conditions of nitrogen starvation. The obvious advantage of the pasteur effect to the micro-organism as less sugar must be degraded to obtain the same amount of ATP when the more efficient aerobic process can be employed.

STUDY OUTLINE

* Pasteur effect can be defined as the slower rate of glucose utilization by a microorganism which is growing aerobically by respiratory metabolism than by the same organism growing anaerobically.

CONCEPT CHECK

What is Pasteur effect? Explain briefly.

18

AMINO ACID BIOSYNTHESIS

Microbial cells require 20 different array of amino acids for protein synthesis. The amino acid biosynthesis may be considered in terms of groups of amino acids formed from common precursors. In addition, as with all small molecules, the precursors arise from catabolic metabolism or an equivalent process. There are six families within the amino acids. All the amino acids are derived from intermediates in glycolysis, the citric acid cycle, or the pentose phosphate pathway. Most bacteria can synthesize all the 20 different amino acids. In the first step, the conversion of nitrogen to glutamine and glutamate takes place. Nitrogen in the form of ammonia (NH_3) is converted into glutamine and glutamate by the help of enzyme glutamine synthetase and glutamate synthetase. This happens in two ways:

  i. α-ketoglutarate – glutamine $\rightarrow$ 2 glutamate
       This reaction takes place in the presence of the enzyme glutamate synthetase.

Table 18.1 Amino acid biosynthetic families grouped by metabolic precursors.

α-ketoglutarate	Oxaloacetate	Phosphoenol pyruvate and erythrose 4-$\textcircled{P}$
Glutamate	Aspartate	
Glutamine	Asparagine	Tryptophan
Proline	Methionine	Phenylalanine
Arginine	Threonine	Tyrosine
	Lysine	
3-Phosphoglycerate	**Pyruvate**	**Ribose 5-phosphate**
Serine	Alanine	
Glycine	Valine	Histidine
Cysteine	Leucine	

ii. In contrast, when α-ketoglutarate reacts with ammonia, it gives glutamate +NADP + ADP + Pi

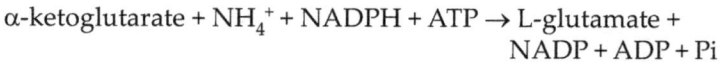

$$\alpha\text{-ketoglutarate} + NH_4^+ + NADPH + ATP \rightarrow L\text{-glutamate} + NADP + ADP + Pi$$

This reaction is mediated by the enzyme glutamate dehydrogenase.

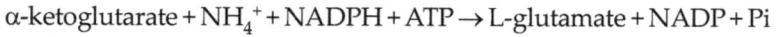

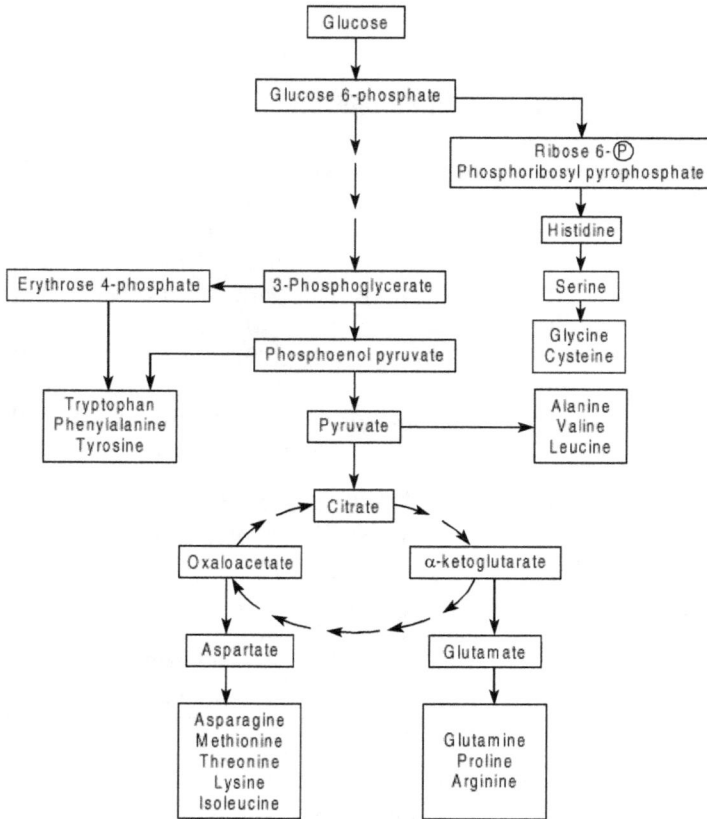

$$\alpha\text{-ketoglutarate} + NH_4^+ + NADPH + ATP \rightarrow L\text{-glutamate} + NADP + Pi$$

Figure 18.1 Overview of amino acid biosynthesis.

THE GLUTAMATE FAMILY

The glutamic acid family includes the linear amino acids glutamine, ornithine, citrulline, and arginine. In addition to linear amino acids,

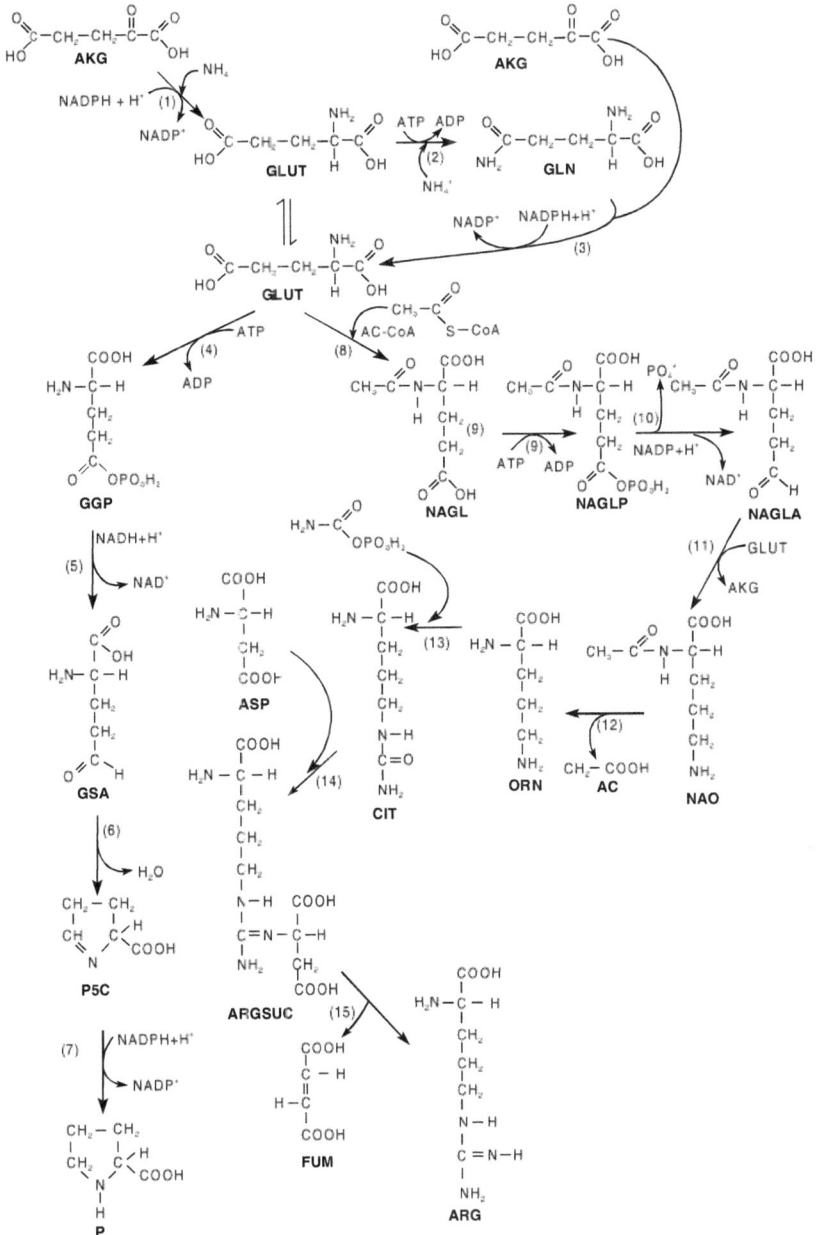

Figure 18.2 Synthesis of the glutamic amino acid family.

the glutamate family includes the cyclic amino acid proline. Glutamic acid (GLUT) arises from α-ketoglutaric acid in one of two ways, either by NH_4 fixation or by combination of L-glutamine with α-ketoglutarate yielding two molecules of glutamate. These two mechanisms operate depending upon the ammonia concentration to which the microbes are exposed. When ammonia concentration reduces below 1 mM, the glutamine synthetase is operative, while at NH_3 concentrations above 1 mM, the dehydrogenase mechanism functions.

THE ASPARTATE FAMILY

The parent amino acid of the aspartate family, aspartic acid arises from the transmination of oxaloacetate and they are further amidated to yield amide asparagine, in a reaction which is analogous to the formation of glutamine and glutamate. The other amino acids belonging to this family are also synthesized by branched pathway. The pathway extends to the synthesis of threonine, the location of branch points leading to lysine, methionine and isoleucine synthesis. The pathway from which lysine is synthesized (sometimes called the diaminopimelic acid or DAP pathway) is characteristic of all prokaryotes, higher plants and most algae. Lysine is synthesized through a different pathway called the α-amino adipic acid or AAA pathway.

Diaminopimelic acid is a component of the peptidoglycan of the cell wall of many eubacteria and dihydrodipicolinic acid is the immediate precursor of dipicolinic acid, a major chemical constituent of endospores that contibutes to their heat stability. In some bacteria the final step of the pathway (methylation) can be catalysed by two distinct enzymes. One enzyme requires folic acid as a cofactor; the other enzyme requires vitamin B_{12}. Some bacteria, for example *E. coli*, can synthesize folic acid, but they are unable to synthesize vitamin B_{12}. The terminal steps in the synthesis of the fifth member of the aspartate family, isoleucine, are catalysed by a series of enzymes that catalyse analogue steps in the biosynthesis of a member of the pyruvate family, valine. Isoleucine biosynthesis will, accordingly, be discussed in the context of valine biosynthesis.

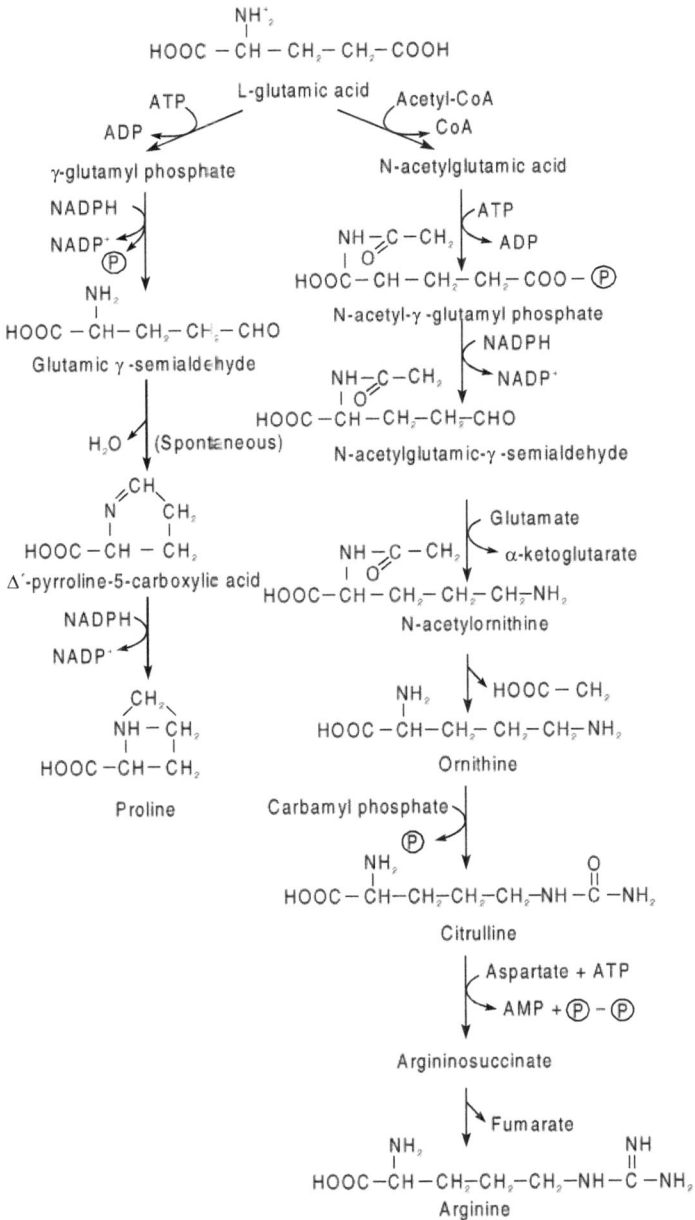

Figure 18.3 Biosynthesis of proline and arginine.

$$
\underset{\text{Oxalacetic acid}}{HOOC - CH_2 - \overset{\overset{\displaystyle O}{\|}}{C} - COOH}
$$

Glutamate

α-ketoglutarate

$$
\underset{\text{Aspartic acid}}{HOOC - CH_2 - \overset{\overset{\displaystyle NH_2}{|}}{CH} - COOH}
$$

ATP   AMP + Ⓟ-Ⓟ

NH₃

$$
\underset{\text{Asparagine}}{NH_2 - \overset{\overset{\displaystyle O}{\|}}{C} - CH_2 - \overset{\overset{\displaystyle NH_2}{|}}{CH} - COOH}
$$

ATP

ADP

β-aspartyl phosphate

NADPH

NADP·   Ⓟ

$$
\underset{\text{Aspartic β-semialdehyde}}{OHC - CH_2 - \overset{\overset{\displaystyle NH_2}{|}}{CH} - COOH} \longrightarrow \longrightarrow \longrightarrow \text{Lysine}
$$

(Figure 18.3)

NADPH

NADP·

$$
\underset{\text{Homoserine}}{HOCH_2 - CH_2 - \overset{\overset{\displaystyle NH_2}{|}}{CH} - COOH} \longrightarrow \longrightarrow \longrightarrow \text{Methionine}
$$

(Figure 18.4)

ATP

ADP

$$
\underset{\text{Homoserine O-phosphate}}{Ⓟ - O - CH_2 - CH_2 - \overset{\overset{\displaystyle NH_2}{|}}{CH} - COOH}
$$

Ⓟ

$$
\underset{\text{Threonine}}{CH_2 - \overset{\overset{\displaystyle OH}{|}}{CH} - \overset{\overset{\displaystyle NH_2}{|}}{CH} - COOH} \longrightarrow \longrightarrow \longrightarrow \text{Isoleucine}
$$

(Figure 18.5)

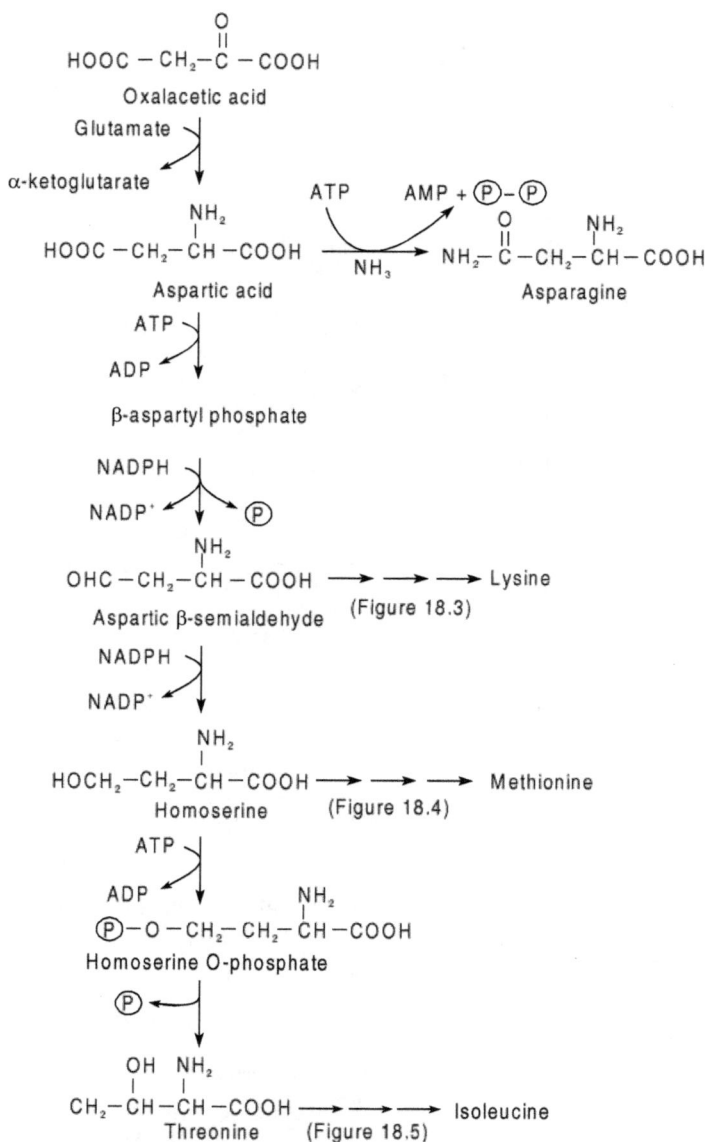

Figure 18.4   Biosynthesis of amino acids of the aspartate family.

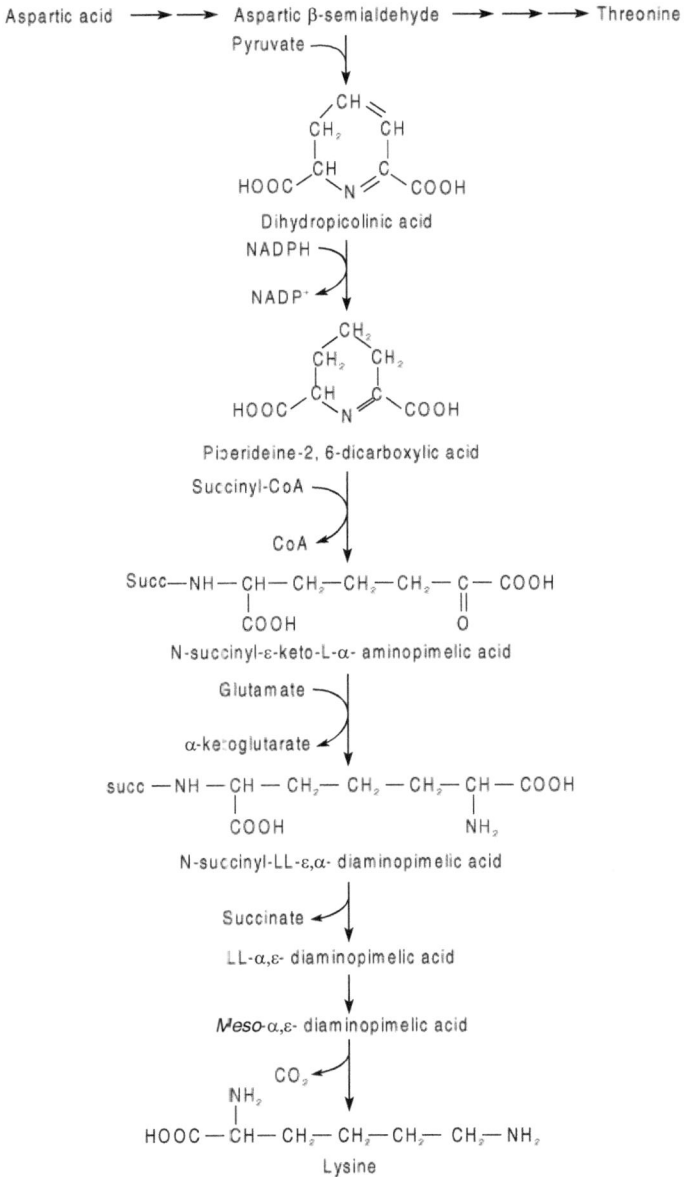

Figure 18.5 The lysine branch of the aspartate pathway (the DAP pathway).

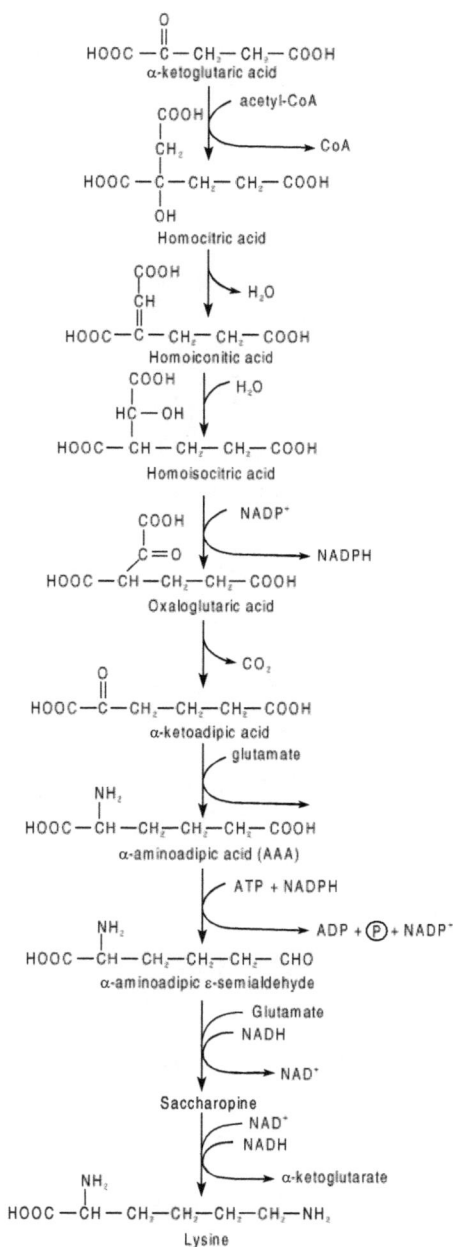

Figure 18.6   The AAA pathway of lysine biosynthesis.

Aspartic acid ⟶ ⟶ ⟶ Homoserine ⟶ ⟶ Threonine

Succinyl-CoA

CoA

NH_2

HOOC—CH—CH_2—CH_2—O—succ
O-succinythomoserine

Cysteine

Succiunate

NH_2 .... NH_2

HOOC—CH—CH_2—CH_2—S—CH_2—CH—COOH
Cystathionine

Pyruvate + NH_3

NH_2

HOOC—CH—CH_2—CH_2—SH
Homocysteine

Methylene FH_4

$[-CH_3]$

Vitamin B_{12} enzyme
+
Methylene FH_4

FH_4

NH_2

HOOC—CH—CH_2—CH_2—S—CH_3
Methionine

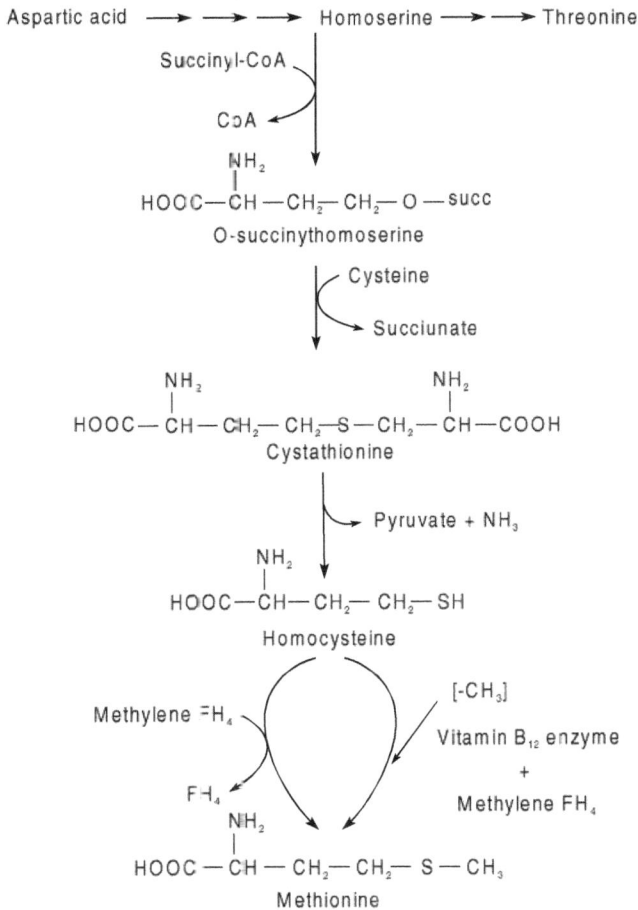

Figure 18.7   The methionine branch of the aspartic acid pathway.

THE AROMATIC FAMILY

Aromatic amino acids include tyrosine, phenylalanine, and tryptophan. The first reaction of this pathway is a condensation between a precursor metabolite from the pentose phosphate cycle, erythrose 4-phosphate, and from the glycolytic pathway, phosphoenol pyruvate. Early steps of this pathway lead to the formation of chorismic acid and prepheric acid, both situated at major metabolic branch points. The aromatic pathway also furnishes, via chorismic acid, P-amino benzoic acid (a precursor of folic acid), p-hydroxybenzoic acid (a precursor of the quinones which are members of certain electron transport chains), and 2, 3-dihydroxybenzoic acid (a component of certain siderophores, which participate in the entry of iron into the cell).

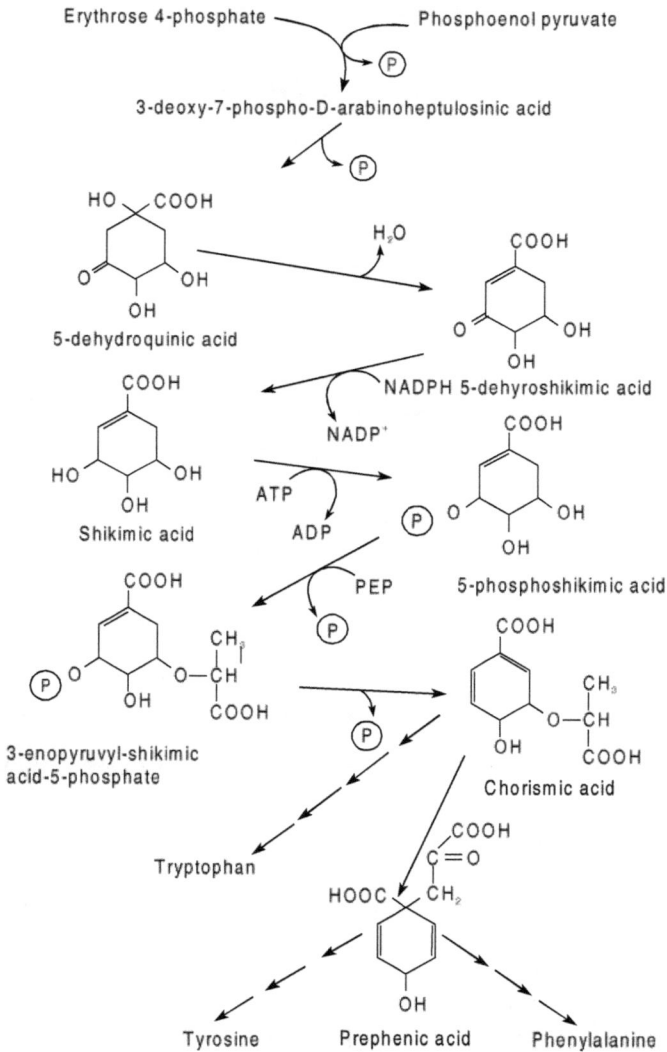

Figure 18.8   Biosynthesis of amino acids of the aromatic family.

THE TRYPTOPHAN BRANCH OF THE AROMATIC ACID PATHWAY

The 3-phosphoglycerate (serine) pathway include serine, cysteine, and glycine. Initially 3-phosphoglycerate (3 PGA) is oxidized to 3-phosphohydroxypyruvate (3 PHP). The 3PHP is then converted to 3-phospho serine (3PS) by amination with glutamate. Serine is then formed by dephosphorylation of 3PS. By transfer of its hydroxymethyl residue to tetrahydrofolate, serine is converted to glycine (GLY). Alternatively serine may react with acetyl-CoA to form O-acetyl serine (OAS). Conversion of OAS to cysteine (cys) results from reaction of OAS with H_2S, often derived from assimilatory sulphate reduction. Serine formation from OAS is accompanied by the release of free acetate.

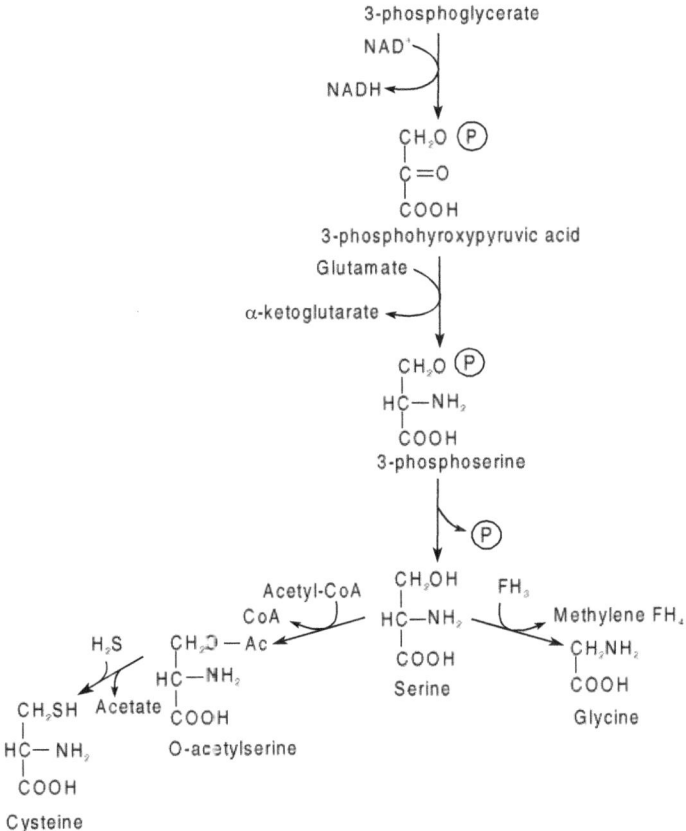

Figure 18.9 Biosynthesis of the amino acids of the serine family.

HISTIDINE SYNTHESIS

In histidine biosynthesis the chain of five carbon atoms in the skeleton of this amino acid is derived from PRPP; two of these atoms contribute to the five-membered imidazole ring and the rest give rise to the three-carbon side chain. The remaining three atoms of the imidazole ring have a curious origin—a C—N fragment is contributed from the purine nucleus of ATP, and the other N atom from glutamine. This utilization of ATP as a donor of two atoms of the purine nucleus is unique. Its physiological rationale lies in the fact that cleavage of the purnine nucleus of ATP leads to the formation of another biosynthetic intermediate, amino imidazole carboxamide ribotide (AICAR) which is itself a precursor of purines. Thus there is an intimate connection between the biosynthesis of histidine and purine.

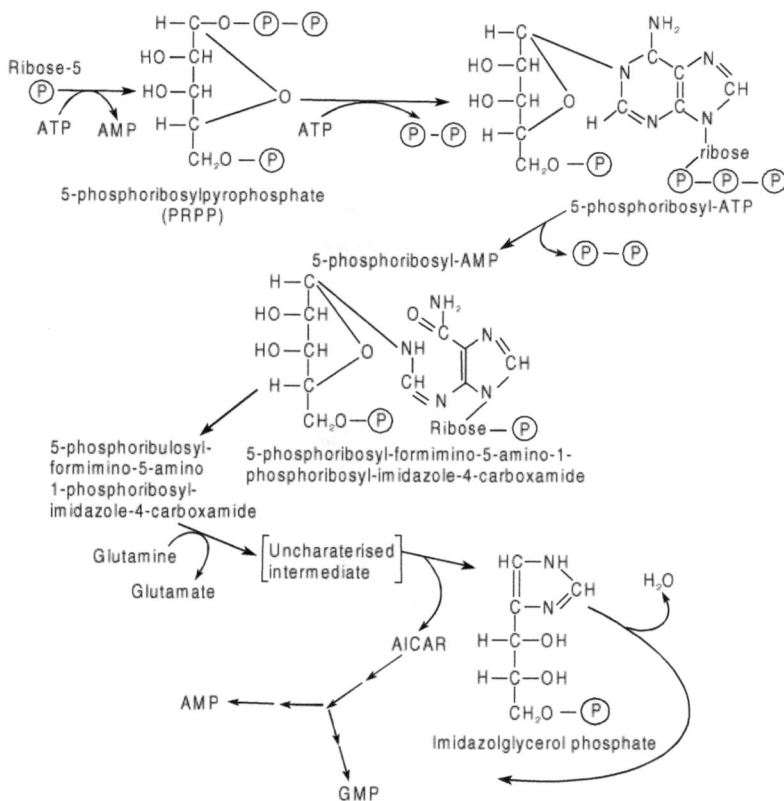

(Contd.)

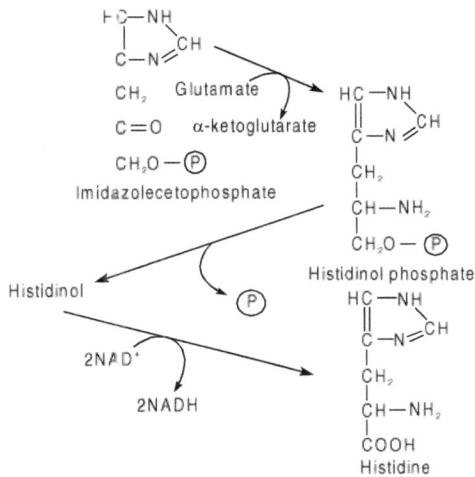

Figure 18.10 The biosynthesis of histidine.

THE UREA CYCLE

Most of the living organisms, during breakdown of amino acids, exerete ammonia as excess nitrogen. Excreted ammonia is very toxic. For this, the ammonia is converted to urea by the system. Accordingly, living organisms are classified as ammonotelic organisms which excrete ammonia, ureotelic organisms which excrete urea and uricotelic organisms which excrete uric acid. Urea is obtained in the liver by the urea cycle. It is then secreted into the bloodstream and sequestered by the kidney for excretion in the urine. The urea cycle was elucidated in *Outline:* 1932 by Hans Krebs and Kurt Hanseleit (the first-known metabolic cycle. Krebs did not elucidate the citric acid cycle until 1937).

In this cycle carbamoyl phosphate synthetase catalyses the condensation and the activation of NH_4^+ and HCO_3^- to form carbamoyl phosphate and for the condensation, energy is taken up by the hydrolysis of 2 ATP molecules. Ornithine transcarbamylase transfers the carbonyl group of carbamoyl phosphate to ornithine, yielding citrulline as a nonstandard amino acid in that it does not occur in proteins. In the next step, the enzyme argininosuccinase catalyses the elimination of arginine from aspartate carbon skeleton forming fumarate. Arginine is urea's intermediate. Next the enzyme arginase catalyses the hydrolysis of arginine to yield urea and regenerate ornithine. It is a nontoxic excretory product of living organisms which needs 2 high energy ATP hydrolysis. Urea cycle is regulated by the substance N-acetyl glutamate. This molecule bonds with the allosteric site of the carbamoyl phosphate synthetase and inhibits the urea cycle.

Structure of N-acetyl glutamate

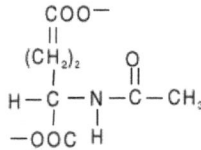

$$2ATP + HCO_3^- + NH_3 \rightarrow H_2N - \overset{O}{\overset{\|}{C}} - OPO_3^{2-} + 2ADP + Pi$$

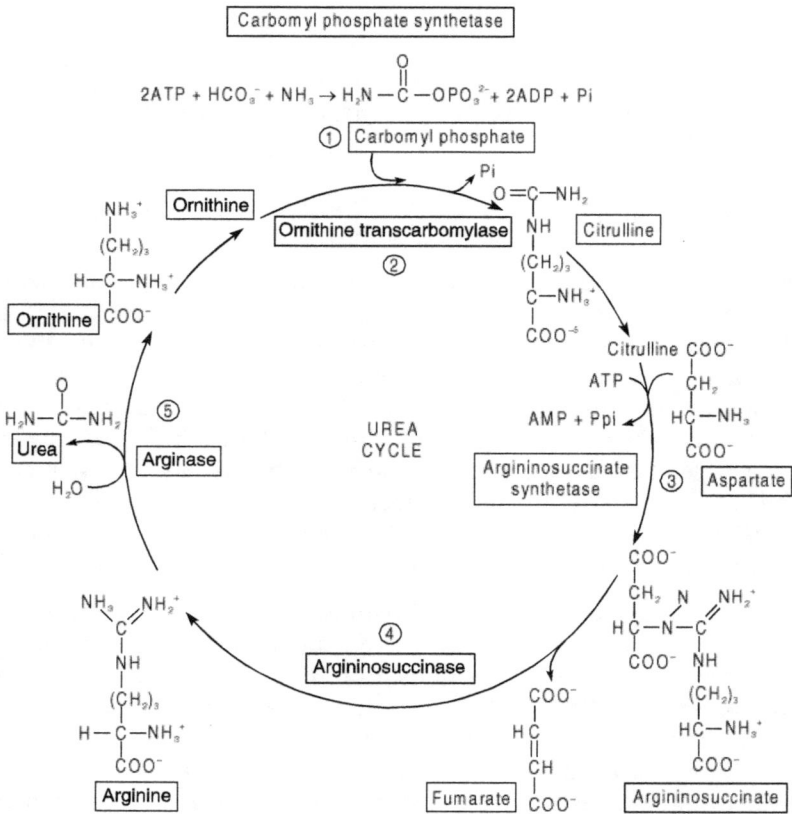

Figure 18.11   Overview of urea cycle.

STUDY OUTLINE

• Amino acids are the building blocks of proteins and may be considered in terms of family groups of amino acids, formed from common precursors.

• All the amino acids are derived from the intermediates in glycolysis, the citric acid cycle, or the pentose phosphate pathway.

• Most of the living organisms, during breakdown of amino acids, excrete ammonia as excess nitrogen. Excreted ammonia is very toxic. This ammonia is converted to urea by the system.

CONCEPT CHECK

1. Write elaborately about the urea cycle.

2. Write about biosynthesis of aromatic amino acids.

3. What is AAA pathway? Explain the steps involved in this pathway.

4. Explain about the biosynthesis of aspartic acid family.

5. How does the biosynthesis of glutamic acid take place?

6. Briefly explain the amino acid biosynthesis of families grouped by metabolic precursors.

CRITICAL THINKING

It has been discussed in this chapter that amino acids are mostly synthesized from the intermediate substances of metabolic pathways. How do microrganisms get their precursors?

19

PROTEIN SYNTHESIS OR TRANSLATION

INTRODUCTION

Protein synthesis is a central physiological process that occurs in all cellular life forms. Proteins mediate all the critical aspects of cellular physiology like the generation of energy, the formation of cell building blocks, the formation of macromolecules, the formation of structures, the uptake and removal of substances, the regulation and coordination of cellular activities, and their own formation. For all of these reasons, it is essential that a microbial physiologist has a fundamental understanding of the process of protein synthesis.

Protein synthesis may be considered on the basis of its components, transcription and translation. Although transcription, the process of forming the message for protein, is not by itself protein synthesis, transcription is essential for the protein synthesis process. Transcription provides the message and order of the amino acids that are linked together to form the protein and also provides molecules of transfer ribonucleic acid (tRNA), the molecule by which activated amino acids may be recognized and added to the growing polypeptide. Furthermore, transcription provides molecules of ribosomal RNA (rRNA) from which the ribosomal particle is made. The intimate interaction between genome, ribosome, and other cell areas are such that it is useful to consider protein synthesis as an example of a multicomponent process that involves the coordinated functioning of various cell components. On the whole "protein synthesis" may be defined as the synthesis of protein using the genetic information in a messenger RNA as a template.

RIBOSOMES

Ribosomes are the sites of protein synthesis. Each ribosome is constructed of two subunits. In prokaryotes, the ribosome subunits are of 30S (Svedberg units) and 50S, yielding intact 70S ribosomes. (The number 30S, 50S, and 70S refer to the sedimentation coefficients of ribosome subunits or intact ribosomes when subjected to centrifugal force in an ultracentrifuge). Each subunit is in itself a ribonucleoprotein complex made up of specific ribosomal RNA and ribosomal proteins. The 30S subunit contains 16S rRNA and about 21 proteins, while the 50S subunit contains 5S and 23S rRNA and about 34 proteins.

BUILDING BLOCKS

Functioning ribosomes depend upon the availability of building blocks that activate amino acids formed at cellular locations other than the genome of the ribosomes. The availability of activated amino acids materially affects the protein synthetic process. Limitations on amino acid availability may result from alterations of the synthetic sequences for particular amino acids, limitations of formation of their activated derivatives, the amino acid adenylates, or limitations of conversion of these derivatives to forms that are recognised by messenger attracted ribosomes (molecules of aminoacyl (tRNA). In addition to those organisms deficient in synthesis of particular amino acids, factors affecting uptake of pre-formed amino acids from the external environment can exert serious effects, not only on protein formation but also on the rate of message synthesis. The formation of mRNA by transcription occurs, and potentially functional ribosomes are of little or no value since they have no "instruction" on how to proceed in the formation of protein. It is only if a code is available that protein can be formed. Formation of the mRNA code is a complicated process that occurs in a substantially different way in prokaryotes than in eukaryotes, although, in principle, the processes are similar.

Protein synthesis involves the following steps:

   i.   Initiation
   ii. Chain elongation
   iii. Termination–release
   iv. Polypeptide folding

INITIATION OF PROTEIN SYNTHESIS

In prokaryotes, the initiation begins with 30S ribosomes and an initiation complex consists of 30S ribosome subunit, mRNA, formyl

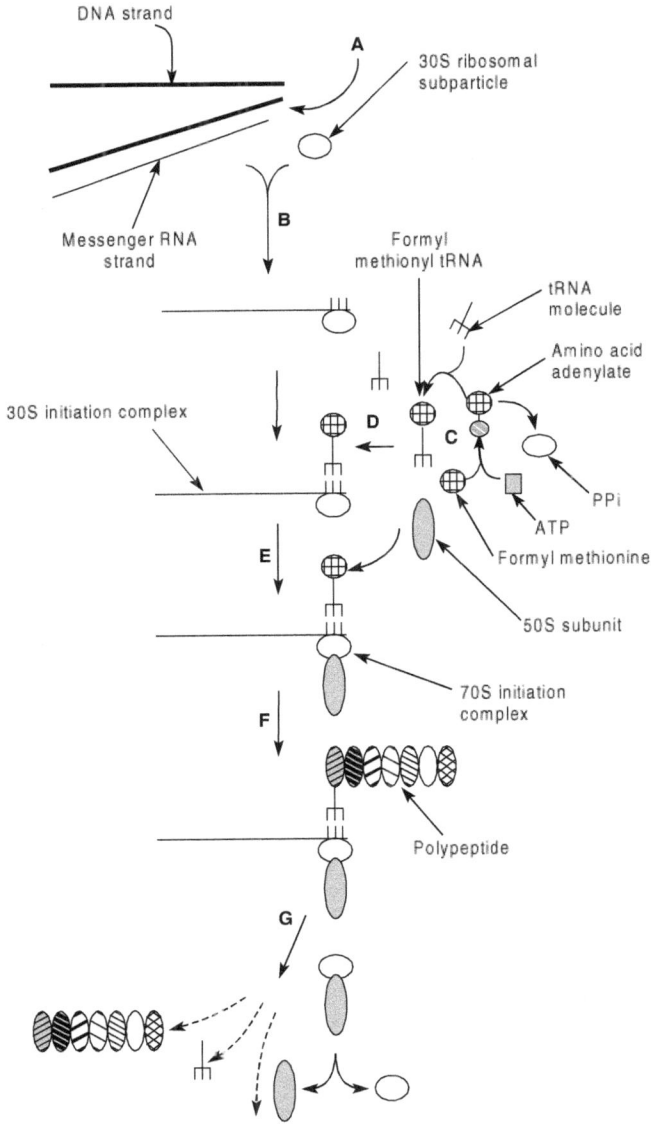

Figure 19.1 An overview of protein synthesis in prokaryotes.

methonine tRNA, and initiation factors, and guanosine 5'-triphosphate (GTP). These molecules comprise the 30S pre-initiation complex, a 50S subunit joins to the 30S subunit to form a 70S initiation complex. This joining requires hydrolysis of the GTP that is present in the 30S preinitiation complex. These sites are called the amino acylsite or A site and the peptidyl or P site; each site consists of a collection of segments of S and L proteins and 23S RNA. The 50S subunit is positioned in the 70S initiation complex such that the fmet tRNA , which has previously bound to the pre-initiation complex, occupies the P site of the 50S subunit. Positioning of tRNA fmet in the P site fixes the position of the anticodon of tRNA fmet such that it can pair with the initiation codon in the mRNA. Thus the reading frame is unambiguously defined upon completion of the 70S initiation complex.

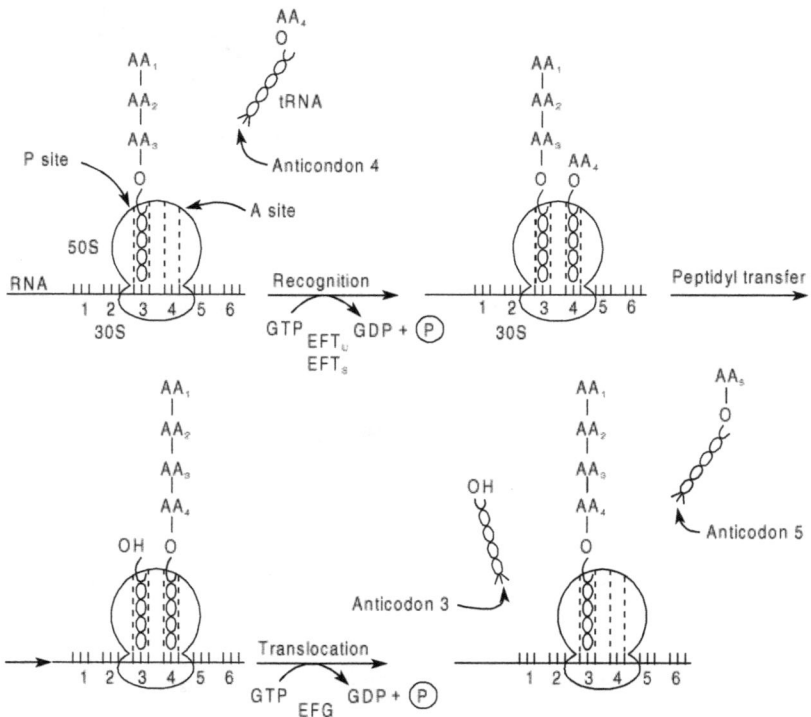

Figure 19.2 The sequence of events in the lengthening of the peptide chain.

ELONGATION

The A site of the 70S initiation complex is available to any tRNA molecule whose anticodon can pair with the codon adjacent to the initiation codon. However, entry to the A site by the tRNA requires a help protein called an elongation factor, Ef (specially Ef-Tu). After occupation of the A site, a peptide bond forms between f-met and the adjacent amino acid formed. Once it was thought that the blockage of the NH_2 group of fmet by the formyl group was responsible for peptide bond formation between the COOH group of fmet and the NH_2 group of the adjacent amino acid. The peptide bond is formed by an unusual enzyme complex called "peptidyl transferase". The active site of peptidyl transferase consists of portions of several proteins of the 50S subunit. As the peptide bond is formed, the fmet is cleared from tRNA fmet in the P site. After the peptide bond formation, an uncharged

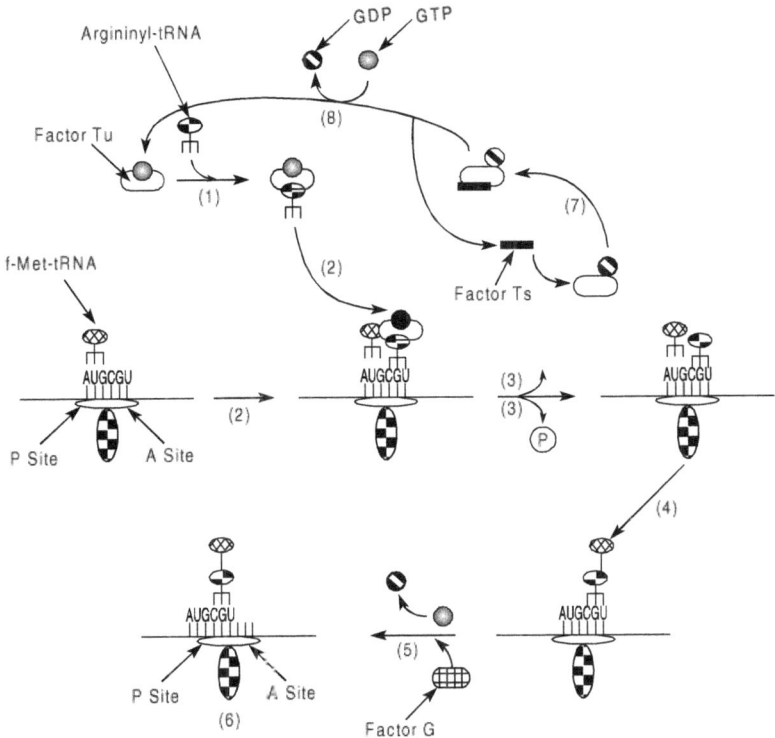

Figure 19.3 The essential aspects of prokaryotic translation.

tRNA occupies the P site and a dipeptidyl tRNA is in the A site. At this point three movements, which together comprise the translocation, occur.

i.   The deacylated tRNA fmet leaves the P site.

ii. The peptidyl tRNA moves from the A site to the P site.

iii. The mRNA moves a distance of three bases in order to position the next codon at the A site.

The translocation requires the presence of another elongation protein Ef-G and hydrolysis of GTP. The movement of the mRNA by three bases is probably dependent on the movement of the tRNA from the A site to the P site and in fact it is likely that mRNA translocation is a consequence of the tRNA motion. The distance of mRNA is only three bases. (If a glycyl frame shift suppressor having an extra base in the anticodon loop is present, the distance of mRNA is four bases instead of three).

After translocation has occurred, the A site is again available to accept a charged tRNA molecule having a correct anticodon. If a $tRNA^{fmet}$ molecule whose anticodon is the same as that of $tRNA^{met}$ molecule, were to enter the A site (because the internal AUG site is present), protein synthesis would stop because a peptide bond cannot form with the blocked NH_2 group of f-met.

TERMINATION

When a chain termination codon is reached, there is no amino acyl tRNA that can fill the A site and chain elongation steps. However the polypeptide chain is still attached to the tRNA occupying the P site. Release of the protein is accomplished by release factor (RF) proteins that in part respond to chain termination codons. There are two such release factors in *E. coli* – RF_1, which recognizes the UAA and UAG codons, and RF_2 which recognizes UAA and UGA. Each releasing factor forms an activated complex with GTP; this complex binds to a termination codon and alters the specificity of peptidyl transferase. In the presence of releasing factors, peptidyl transferase catalyses the reaction of the bound peptidyl moiety with water rather than with the free amino acyl tRNA. Thus the polypeptide chains, which have been held in the ribosome solely by the interaction with the tRNA in the P site is released from the ribosome. Finally the 70S ribosome dissociates into 30S and 50S subunits and the system is now ready to start synthesis of a second chain.

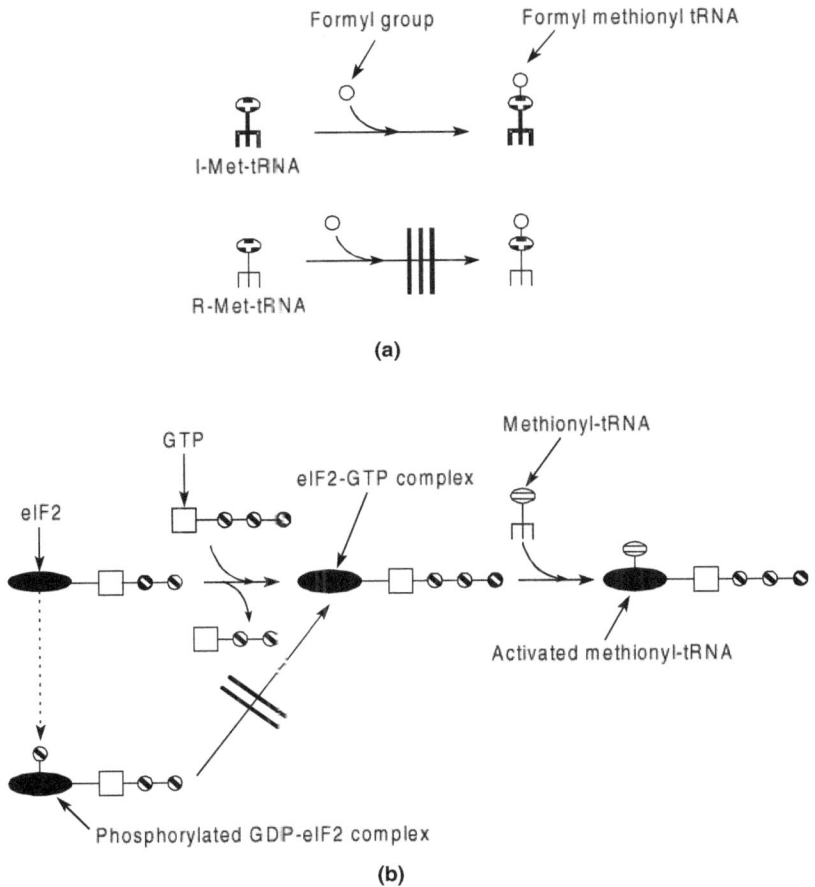

(a)

(b)

Figure 19.4 Regulation of translation initiation in prokaryotes (a) and eukaryotes (b).

POST-TRANSLATIONAL MODIFICATION OR PROTEIN PROCESSING

After protein synthesis the synthesized polypeptide chain has to be processed and folded by some modifications. During or after its synthesis, the polypeptide progressively assumes its native conformation, with the formation of appropriate hydrogen bonds and vander Waals, ionic, and hydrophobic interactions. By this the linear, or one-dimensional genetic message in the mRNA is corrected into three-dimensional structure of protein. Some newly made proteins, both prokaryotic and eukaryotic, do not attain their final biologically active conformation until they have been altered by one or more processing reactions called "Post translational modification."

Various modifications are involved depending on the nature of protein.

Amino Terminal and Carboxy Terminal Modification

The first amino acid residue inserted in all polypeptide chains is methionine (in the case of eukaryote) or formyl methionine (in the case of prokaryotes). This N-terminal amino group residue is removed by N-acylation after translation. Carboxy terminal residues are also sometimes modified.

Loss of Signal Sequence

Signal sequences are short sequences of amino acid residues, which direct a protein to its appropriate location in the cell. The signal sequence is 15–30 residues at the amino terminal end of some proteins and plays a role in directing the protein to its ultimate destination in the cell. Such signal sequences are ultimately removed by specific peptidases.

Modifications of Individual Amino Acids

The hydroxyl group of certain serine, threonine and tyrosine residues of some proteins are enzymatically phosphorylated by ATP.

This phosphorylation contributes negative charges to those polypeptide chains. The functional significance of these modifications varies from one protein to the next.

For example, the milk protein casein has many phosphoserine groups that bind Ca^{2+}. Calcium, phosphate and amino acids are all valuable for suckling young. So casein effectively provides three essential nutrients. So the phosphorylation and dephosphorylation cycles regulate the activity of many enzyme and regulatory proteins.

$$COO^-$$
$$H_3N-C-H$$
$$CH_2$$

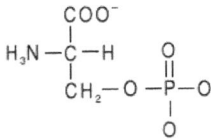

Phosphoserine   Phosphothreonine   Phosphotyrosine

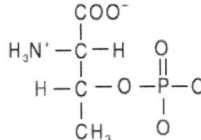

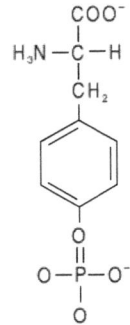

Addition of Carboxyl Group

Extra carboxyl group may be added to glutamate residues of some proteins. For example, the blood clotting factor (protein prothrombin) contains amino terminal region introduced by an enzyme that requires vitamin K. These carboxyl groups bind with Ca^{2+}, which is required to initate the clotting mechanism.

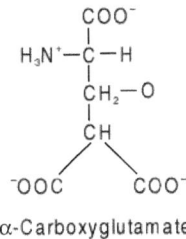

α-Carboxyglutamate

In some proteins, the carboxyl groups of some glutamate residues undergo methylation, removing their negative charge.

Attachment of Carbohydrate Side Chains

In the case of glycoproteins the carbohydrate side chains are covalently attached after translation. They may be attached enzymatically with asparagine (N-linked oligosaccharides) in others to serine or threonine residues (O-linked oligosaccharide). These glycoproteins function extracellularly, as well as act as "lubricating proteoglycans" that coat mucous membranes that contain oligosaccharide side chains.

Addition of Isoprenyl Group

A number of eukaryotic proteins are modified by the additive of groups derived from isoprene (isoprenyl group). A thioether bond is formed between the isoprenyl group and cysteine residues of the protein.

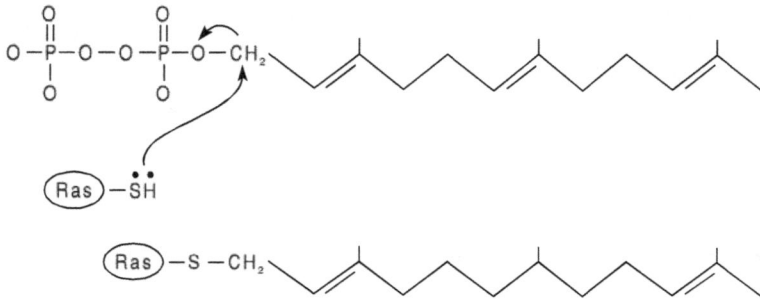

The isoprenyl groups are derived from phosphorylated intermediates of the cholesterol biosynthetic pathway, such as farnesyl pyrophosphate. Proteins modified in this way include the Ras protein, product of the ras oncogenes and protooncogene, and G proteins and lamins found in the nuclear membrane. The transforming (carcinogenic) activity of the ras oncongenes is lost when isoprenylation of the ras protein is blocked (a finding that has stimulated interest in identifying inhibitors of this post-translational modification pathway for use in cancer chemotherapy).

Addition of Prosthetic Groups

Many prokaryotic and eukaryotic proteins require for their activity covalently bound prosthetic groups. Two examples are the biotin molecules of acetyl-CoA carboxylase and the heme group of hemoglobin or cytochrome c.

Proteolytic Processing

Many proteins are initially synthesized as large inactive precursor polypeptides that are proteolytically trimmed to form their smaller active forms, e.g. proinsulin, some viral proteins and proteins such as chymotrypsinogen and trypsinogen.

Formation of Disulphide Bridges

After folding into their native conformations, some proteins form intra-chain or inter-chain disulphide bridges between cysteine residues. In eukaryotes, disulphide bonds are common in proteins to be exported from cells. Cross-links formed in this way help to protect the natural conformation of the protein molecule from denaturation in the extracellular environment, which can differ greatly from intracellular conditions and is generally oxidizing.

ANTIBIOTICS AND TOXINS INHIBITING PROTEIN SYNTHESIS

Puromycin

The mold *Streptomyces alboniger* (source) is the best inhibitory antibiotic of protein synthesis.

Mode of action Its structure is very similar to the 3' end of an amino-acyl–tRNA, enabling it to bind to the ribosomal A site and participate in peptide bond formation producing peptidyl puromycin. However, puromycin resembles only the 3' end of the tRNA, it does not engage in translocation and dissociates from the ribosome shortly after it is linked to the carboxy terminal of the peptide. This will prematurely terminate polypeptide synthesis.

Tetracycline

These block the A site on the ribosome, and thus prevent the binding of amino acyl tRNA.

Chloramphenicol

This inhibits protein synthesis by bacterial (mitochondria and chloroplast) ribosomes by blocking peptidyl transferase; it does not affect cytosolic protein synthesis in eukaryotes.

Cycloheximide

This blocks the peptide transferase of 80S eukaryotic ribosome but not that of the prokaryotic ribosome.

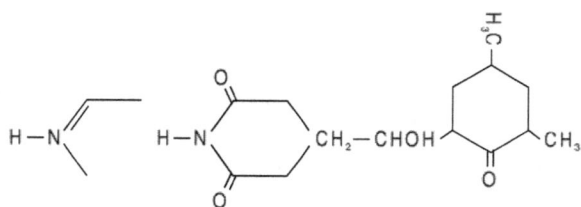

Structure of cycloheximide

Streptomycin

A basic trisaccharide causes misreading of the genetic code (in bacteria) at relatively low concentration and inhibits initiation at low concentration.

Toxins

Diptheria toxin (Mr. 58, 330) catalyses the ADP ribosylation of a dipthamide (a modified histidine) residue of eukaryotic elongation factor EF_2 thereby inactivating it.

Ricin (Mr. 29, 895), an extremely toxic protein of the bean inactivates the 60S subunit of eukaryotic ribosomes by depurinating a specific adenosine in 23S rRNA.

PROTEIN CATABOLISM

Catabolism of protein is essential for the catabolism of amino acids. Amino acids are highly reactive and undergo rapid changes. They may be utilized to form a new protein or they replace amino acids that are lost. Amino acids contribute to the formation of enzymes, pigments, toxins, antibiotics, some important proteins, and some surface antigens. In general, amino acids catabolize with the separation of amino group, leaving behind the carbon skeleton, which becomes a keto acid. The ammonia enters the ammonia pool where it is utilized in various anabolic and catabolic reactions. The majority of the α keto acid enters the carbohydrate pool and some participate in the formation of fatty acids.

The catabolic pathway of amino acids involves the following reactions.

   i.   Removal of amino group

   ii.   Fate of amino group

   iii.   Fate of carbon skeleton

Removal of Amino Group from Amino Acid

Removal of amino group is accomplished by way of: (a) oxidative deamination (b) Transamination (c) transdeamination.

Oxidative deamination In oxidative deamination, oxidation of amino acid takes place by the help of the enzyme amino acid oxidase or dehydrogenase (Both L and D amino acid oxidase) which is mediated by FAD. Oxidative deamination takes places in two stages (i) Dehydrogenation of amino acid (or oxidation) and (ii) Hydrolysis of amino acid followed by liberation of ammonia and keto acid. First the amino acid is converted into imino acid.

e.g. $CH_3CHNH_2COOH + FAD \xrightarrow[\substack{\text{Amino acid}\\\text{oxidase}}]{2H^+} CH_3 = NHCOOH + FADH_2$ (Imino acid)

1. $CH_3C = NHCOOH - \xrightarrow{H_2O} CH_3\overset{\overset{O}{\|}}{C} - COOH + NH_3$
                                Pyruvic acid + ammonia
                                  keto acid

FADH2 is reoxidized to forms

$$FADH_2 + O_2 \longrightarrow H_2O_2 + FAD$$

The overall net reaction is as follows.

$$\underset{\text{Amino acid}}{R-\overset{\overset{H}{|}}{\underset{|}{C}}-NH_2} \xrightarrow{-2H^+} \underset{\text{Imino acid}}{R-\underset{|}{C}=NH} \xrightarrow{H_2O} \underset{\text{Keto acid}}{R-\underset{|}{C}=O} + NH_3 \text{(ammonia)}$$

                    COOH   COOH   COOH

2. $HOOC-CH_2-CHNH_2-COOH \xrightarrow{-2H} HOOC-CH_2C = NH-COOH \xrightarrow{H_2O}$
       Aspartic acid

$$HOOC - CH_2 - \overset{\overset{O}{\|}}{C} - COOH + NH_3$$
Oxaloacetic acid

3.   Glutamic acid $\longrightarrow$ α-ketoglutaric acid + NH_3

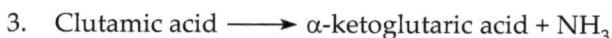

Transamination   The next step is transamination, i.e., removal of an amino group from one amino acid which is coupled with one keto acid to form a new ketoacid and a new amino acid. This reaction is meditated by transaminase which meditates the transfer of amino group from glutamate which is coupled with oxaloacetate (keto acid) which forms α-ketoglutarate and aspartate.

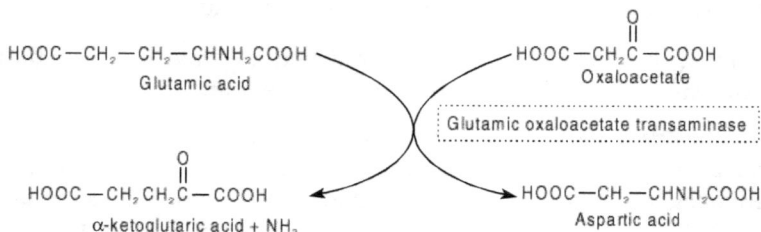

$$HOOC-CH_2-CH_2-CHNH_2COOH \qquad HOOC-CH_2\overset{\overset{O}{\|}}{C}-COOH$$

Glutamic acid   Oxaloacetate

Glutamic oxaloacetate transaminase

$$HOOC-CH_2CH_2\overset{\overset{O}{\|}}{C}-COOH \qquad HOOC-CH_2-CHNH_2COOH$$

α-ketoglutaric acid + NH_3 Aspartic acid

Transdeamination   Transdeamination is the final step, and is a combination of transamination and deamination. In this process the amino group is removed from one amino acid which combines with another keto acid forming an amino acid.

e.g., Glutamic acid

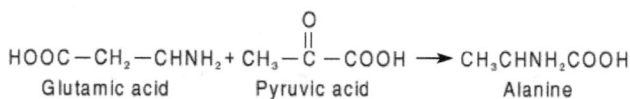

$$HOOC-CH_2-CHNH_2+ CH_3-\overset{\overset{O}{\|}}{C}-COOH \longrightarrow CH_3CHNH_2COOH$$

Glutamic acid   Pyruvic acid   Alanine

Alanine further undergoes oxidative deamination to give

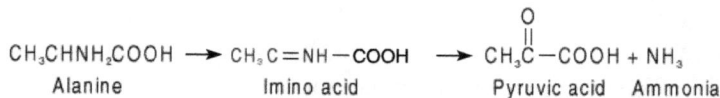

$$CH_3CHNH_2COOH \longrightarrow CH_3C=NH-COOH \longrightarrow CH_3\overset{\overset{O}{\|}}{C}-COOH + NH_3$$

Alanine   Imino acid   Pyruvic acid   Ammonia

Fate of Amino Group

The second process in the catabolism of amino acids is the process deciding the fate of the amino group. In this process, ammonia released by deamination of amino acids, enters the general ammonia pool where it may be utilized for either anabolic or catabolic purposes. This includes (i) synthetic pathway (ii) glutamine pathway and (iii) formation of urea.

Synthetic pathway   In synthetic pathway α-keto amino acids derived from carbohydrates may be reductively aminated to form new amino acids by the reversal of transdeamination reaction. Ammonia may be utilized for the synthesis of purines and pyrimidines.

Glutamine pathway If ammonia is largely accumulated in a cell, it is toxic to the cell. To detoxify this ammonia, it is combined with glutamic acid and forms glutamine.

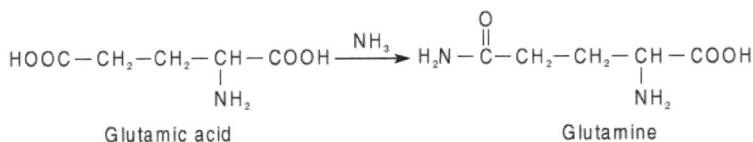

$$\text{HOOC}-CH_2-CH_2-\underset{\underset{NH_2}{|}}{CH}-COOH \xrightarrow{NH_3} H_2N-\overset{\overset{O}{\|}}{C}-CH_2-CH_2-\underset{\underset{NH_2}{|}}{CH}-COOH$$

Glutamic acid   Glutamine

Formation of urea After deamination, the ammonia is converted to urea if it would enter the ornithine cycle.

Fate of Carbon Skeleton

The liberation of ammonia from an amino acid and the presence of carbon skeleton is considered as α-keto acid. The α-keto acids are used by the following ways.

i. The α-keto acids are re-used for the synthesis of amino acid by reverse deamination process.

ii. The α-keto acids are utilized in glucogenic pathway for the carbohydrates and some α-keto acids are used as intermediates of carbohydrate metabolism.

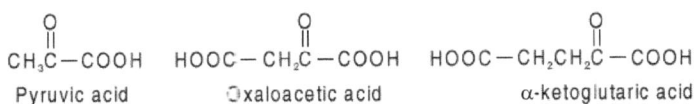

$$\underset{\text{Pyruvic acid}}{CH_3\overset{\overset{O}{\|}}{C}-COOH} \qquad \underset{\text{Oxaloacetic acid}}{HOOC-CH_2\overset{\overset{O}{\|}}{C}-COOH} \qquad \underset{\text{α-ketoglutaric acid}}{HOOC-CH_2CH_2\overset{\overset{O}{\|}}{C}-COOH}$$

But in ketogenic pathway, some α-keto acids are used for the biosynthesis of fatty acids.

STUDY OUTLINE

*   Alignment of amino acids correspond to the message in the mRNA which leads to the formation of protein. This is termed as translation.

*   Protein synthesis is an essential event in cellular metabolism. Proteins are needed by the cell for various metabolic processes.

*   The basic requirements for protein synthesis are mRNA, tRNA ribosomes, enzymes and building blocks clearly known as amino acids.

- Protein synthesis generally consists of three steps namely initiation, elongation and chain termination.
- For the initial step the ribosomes should be in dissociated form.
- Peptide bond formation in protein synthesis is initiated by peptidase enzyme.
- Protein synthesis could be blocked by some drugs like tetracycline, chloramphenicol and cycloheximide.
- After the protein is synthesized, it must be processed for its native conformation. Various steps are involved in the processing of protein.
- The processing of protein depends upon the type of system, it synthesizes. Different systems have different processing mechanisms.

CONCEPT CHECK

1. Write about the basic requirements for protein synthesis and their role in protein synthesis process.

2. What is protein synthesis? Explain the events in protein synthesis.

3. Outline the name of the drugs and their mode of action in blocking protein synthesis.

4. How are the proteins processed after their translation?

   or

   Explain about post-translational modifications.

5. Briefly explain about the catabolism of protein.

CRITICAL THINKING

Is it possible to block protein synthesis using anti-sense RNA technology?

20

BIOSYNTHESIS OF MACROMOLECULES

INTRODUCTION

Each and every constituent of a cell is made up of macromolecules. Macromolecules include proteins, amino acids, Nucleic acids, carbohydrates, and complex lipids. Micromolecules are composed of lower molecular weight monomer molecules which are termed as building blocks. Nucleic acids are composed of nucleotides. Proteins are composed of amino acids, carbohydrates (polysaccharides) are composed of simple sugars (monosaccharides), complex lipids are composed of fatty acids, polyalcohols, simple sugars, amines and amino acids. Nearly 70 different building blocks are needed for the composing of four classes of macromolecules. Each macromolecule synthesis consists of various steps which need compounds for catalytic roles. These include about 20 coenzymes and electron carriers. About 150 different small molecules are required to produce a new cell. These small molecules are in turn synthesized from the 12 precursor metabolites formed in the course of catabolism by heterotrophs or of CO_2 assimilation by autotrophs.

BIOSYNTHESIS OF PURINE AND PYRIMIDINE NUCLEOTIDES

Purines and pyrimidines are the precursors for the nucleic acid biosynthesis. These molecules build the nucleic acid blocks. In this section, we will discuss how they are assembled into macromolecules and how they form the cellular structures. Both purines and pyrimidines can exist in two forms—nucleoside and nucleotide. Nucleoside is nothing but the attachment of pyrimidine bases through nitrogen atoms to a pentose sugar. But a nucleotide is the attachment of nucleoside with a phosphate group (to distinguish between the base and pentose moieties of a nucleoside, the position on the pentose are assigned a prime following the number).

Table 20.1   Name and composition of nucleoside triphosphate.

Base		Ribonucleosides		2-deoxyribnucleosides	
Name	Base structure	Pentose structure	Name	Pentose structure	Name
Purines					
Adenine	(structure)		Adenosine		2-Dexoyadenosine
Guanine	(structure)	Ribose (structure)	Guanosine	2-deoxyribose (structure)	2-Deoxygunosine
Pyrimidines					
Uracil	(structure)		Uridine		2-Deoxyuridine
Cytosine	(structure)		Cytidine		2-Deoxycytidine
Thymine	(structure)		Ribothymidine		2-Deoxythymidine

Nucleotides are symbolized by letters, A, G, U, C, or T, which indicates the purine and pyrimidines, and MP, DP, and TP indicate that the mono, di, triphosphates are attached with the nucleoside. The lower case "d" indicates the deoxy sugar. For example TDP indicates thymidine diphosphate. However, dCTP indicates the deoxycytidine triphosphate. The two purines (dATP and dGTP) and two pyrimidines (dTTP and dCTP) are the precursors for nucleic acid biosynthesis.

For ribonucleic acid biosynthesis, instead of thymidine, uridine triphosphate is used. Some nucleotides can act as activators and thus play dual roles. In the formation of nucleotides, first ribonucleotides are formed. The reduction of ribonucleotides leads to the formation of deoxyribonucleotides.

Nucleotides serve the additional function of being precursors of uric acid, the end product of protein nitrogen catabolism. The three processes that contribute to purine nucleotide biosynthesis, listed in decreasing order are (1) synthesis from amphibolic intermediates (synthesis denovo) (2) phosphoribosylation of purines and (3) phosphorylation of purine nucleosides. The sources of the nitrogen and carbon atoms of the purines are illustrated in figure 20.1.

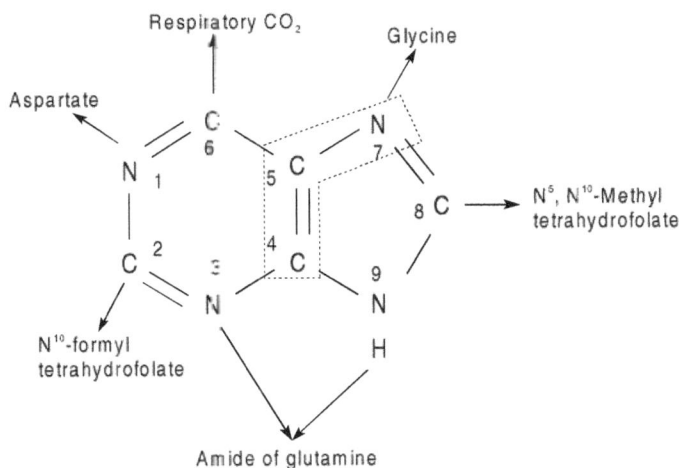

Figure 20.1 Source of nitrogen and carbon atoms of purines.

Nucleotides can be synthesized by either of the two ways:

1. De novo pathway
2. Salvage pathway

DE NOVO PATHWAY

The synthesis of nucleotides by forming a parent molecule IMP is called de novo synthesis. IMP is synthesized from ribose 5-phosphate.

The IMP is the parent nucleotide from which both AMP and GMP are formed. Synthesis of inosine monophosphate (IMP) starts with the amphibolic intermediate α-D-ribose 5-phosphate and involves a linear sequence of 11 reactions. The pathway has branches. One leading from IMP to AMP, the other from IMP to GMP. The first step for IMP biosynthesis starts with the amphibolic intermediate derived from ribose 5-phosphate moiety of all ribonucleotides is derived from 5-phosphoribosyl which in turn is synthesized from ribose 5-phosphate (a precursor metabolite generated in the pentose phosphate pathway) and ATP.

$$\text{Ribose 5-phosphate} + \text{ATP} \xrightarrow[\text{synthetase}]{\text{PRPP}} \text{PRPP} + \text{AMP}$$

In purine nucleotide biosynthesis, PRPP is the starting point. By adding the amino groups and small carbon-containing groups, the nine-numbered ring is synthesized. Three processes contribute to the purine nucleotide biosynthesis. They are (1) synthesis from amphibolic intermediates (synthesis de novo) (2) phosphoribosylation of purines and (3) phosphorylation of purine nucleotides. In prokaryotes each reaction is catalysed by different polypeptides. But in the case of eukaryotes, the enzymes are polypeptides with multiple catalytic activities and adjacent catalytic sites.

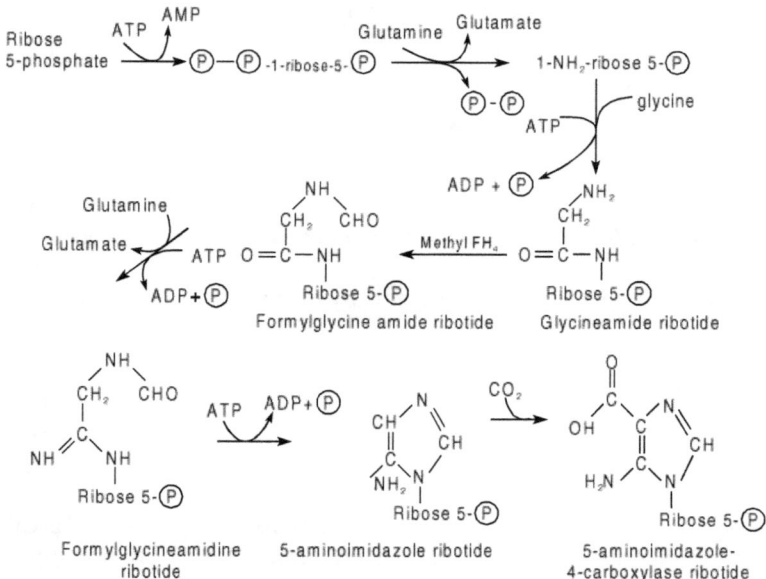

Formylglycineamidine ribotide   5-aminoimidazole ribotide   5-aminoimidazole-4-carboxylase ribotide

(Contd.)

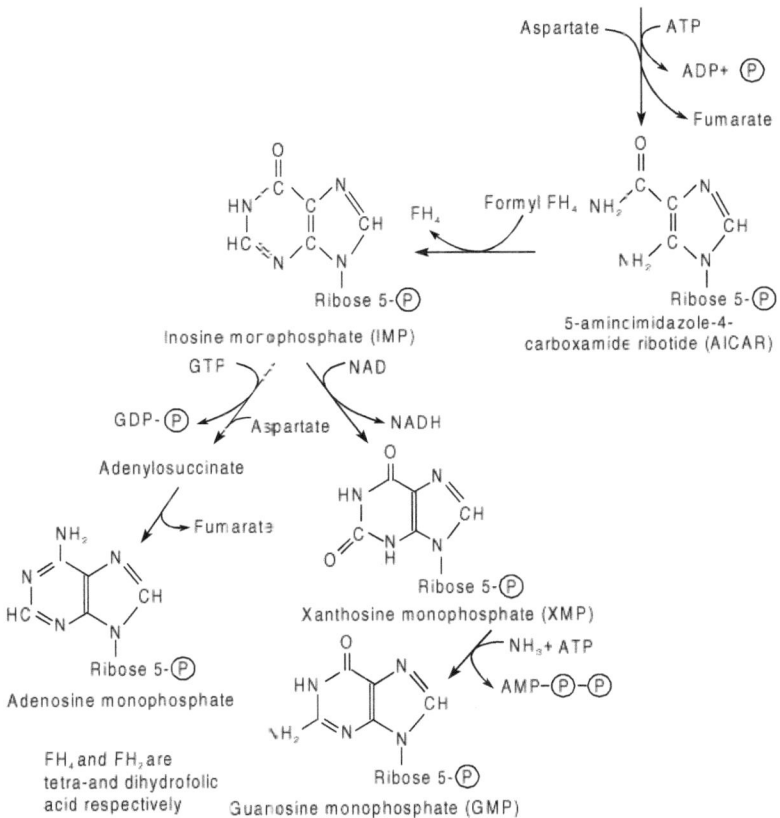

Figure 20.2 The biosynthesis of purine nucleotides.

With the single exception of CTP, all nucleotide triphosphates are synthesized from the corresponding nucleoside monophosphates.

The reactions of IMP to GMP and AMP are irreversible. So the recycling of GMP and AMP to IMP takes place by the ancillary pathways. Thus external sources of guanine and adenine satisfy the cell requirements. The pathways between ATP and aminoimidazole carboxymide ribotide (AICAR) is also common to the pathway by which the amino acid, histidine is synthesized. The two purine ribonucleoside monophosphates, AMP and GMP are the precursors of the four essential ribonucleoside tri phosphates like ATP, GTP, UTP, and CTP.

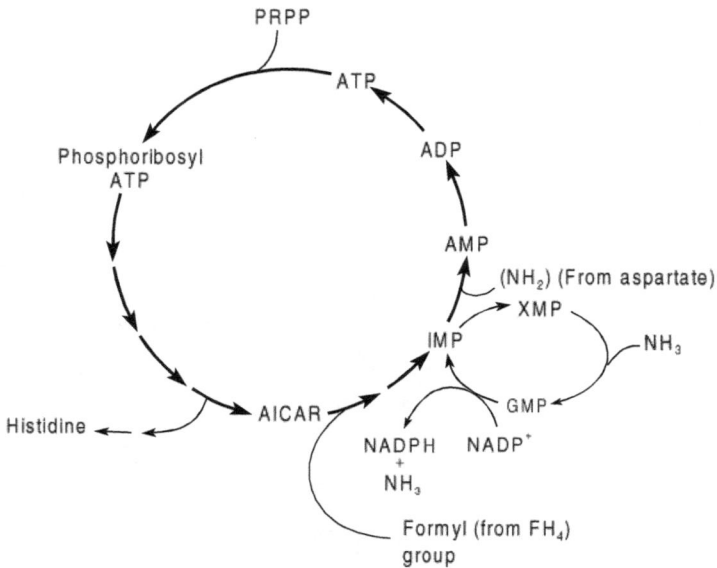

Figure 20.3 Interconversion pathways between GMP and AMP and the relation of one of these to the biosynthesis of the amino acid, histidine.

Synthesis of the 2-deoxyribonucleotides

The four deoxyribonucleotides are synthesized from ribonucleotides. Three nucleotides (dATP, dGTP, dCTP) are formed by the reduction of ribonucleotides which is regulated by a single, highly regulated enzyme complex. In some bacteria like *E. coli* the reduction takes place at the level of nucleoside diphosphate. But in lactic acid bacteria, the reduction takes place at the level of nucleoside triphosphate. The enzyme nucleoside diphosphokinase_converts the nucleoside diphosphate formed from the reduction reaction to nucleoside triphosphate. Next is the synthesis of dTTP which requires dUTP. dUTP is not normally a precursor for the dTTP. They are intermediate products. This dUTP is formed from dCTP by the deamination of cytosine and followed by nucleoside diphosphokinase activity. dIMP is formed by the methylation of dUTP and before this they are converted to dUMP by the action of the enzyme pyrophosphate and after formation of dTTP, they are returned to the triphosphate level by two kinase reactions. This reaction burns with the help of the fuel ATP.

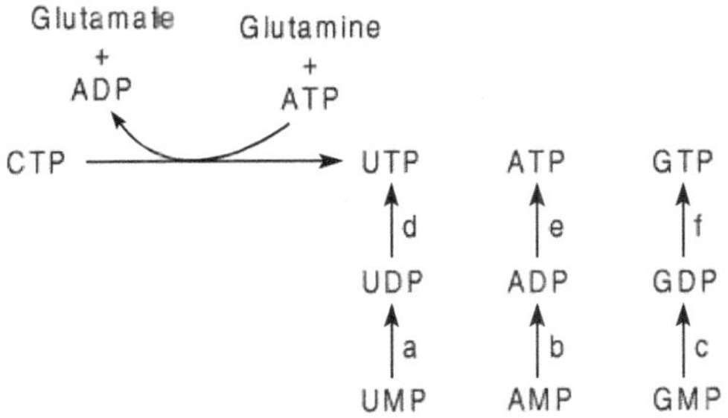

Glutamate   Glutamine
     +   +
   ADP   ATP

CTP ───────────⟶ UTP   ATP   GTP
                      ↑d   ↑e   ↑f
                     UDP   ADP   GDP
                      ↑a   ↑b   ↑c
                     UMP   AMP   GMP

Figure 20.4   Formation of ribonucleoside triphosphate from UMP.

Each reaction is catalysed by a specific kinase.

Biosynthesis of Pyrimidine Nucleotides

The purine and pyrimidine nucleotides have several common precursors—PRPP, Glutamine, CO_2, aspartate, and for thymine nucleotides, tetahydrofolate.

The synthesis of pyrimidine starts with the precursors glutamate, ATP, and CO_2, of carbamoyl phosphate and this reaction is carried out by the enzyme carbamoyl phosphate synthase II. In the next step condensation of carbamoyl phosphate with aspartate forms carbamoyl aspartate in a reaction catalysed by aspartate transcarbamoylase. The carbamoyl aspartic acid ring is closed by loss of water, catalysed by dihydroorotate, which forms dihydroorotic acid.

In the next step, the hydrogen from C-5 and C-6 is removed by NAD^+ and forms $NADH + H^+$, forming orotic acid. This reaction is mediated by the enzyme dihydroorotate dehydrogenase. Then the phosphate group is attached, and this phosphate group is added by ribose phosphate moiety from PRPP. The product formed here is orotidine monophosphate, and this reaction is mediated by the enzyme phosphoribosyl transferase. The product orotidylate is then decarboxylated and forms uridine monophosphate which is the first pyrimidine ribonucleotide. The pyrimidine nucleotide in this step is phosphorylated with the help of ATP and forms UTP. The UTP under deamination by glutamine forms CTP. In the final methylation of dUMP at C-5 by N^5, N^{10}, methylene tetrahydrofolate, catalysed by thymidylate synthase, forms thymidine monophosphate (TMP).

Drugs inhibition on nucleotide biosynthesis   The two carbons are inserted from N^5, N^{10}-methenyl and N^{10} formyl tetrahydrofolate. Drugs like azaserine, diazanorleucine and 6-mercaptopurine and mycophenolic acid inhibit the various reactions in nucleotide biosynthesis (purine). The inhibition of formation of tetrahydrofolate compound can block purine synthesis.

SALVAGE PATHWAY OF PURINE AND PYRIMIDINE NUCLEOTIDES

Some bacteria carry out some special type of nucleotide synthesis. These organisms are also able to utilize purines and pyrimidines in the form of free bases as well as nucleotides, when these compounds are supplied in the medium. This reaction requires less energy than does de novo synthesis. The most important mechanism involved in this salvage pathway is phosphoribosylation of a free purine (Pu) by PRPP, forming a purine-5-mononucleotide (Pu-RP).

$$Pu + PP–RP \longrightarrow Pu-RP + PPi$$

The pathway by which free bases and nucleosides are utilized when supplied exogenously have been called salvage pathway.

The PRPP-dependent phosphoribosylation of purines is catalysed by adenine phosphoribosyl transferase and hypoxanthine–guanine phosphoribosyl transferase. The former enzyme converts the adenine to AMP and the latter enzyme converts hypoxanthine or guanine to IMP or GMP.

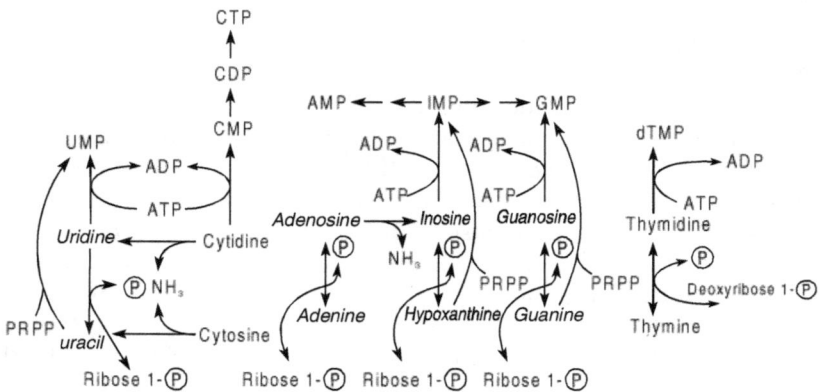

Figure 20.5   Pathways in enteric bacteria for the utilization of exogenous sources of purine and pyrimidine nucleotides.

A second salvage mechanism involves direct phosphorylation of a purine ribonucleoside (PuR) by ATP.

$$PuR + ATP \longrightarrow PuR\text{-}P + ADP$$

Adenosine kinase catalyses phosphorylation of adenosine to AMP or of deoxyadenosine to dAMP. Deoxycytidine kinase phosphorylates deoxycytidine, deoxyadenosine, and 2-deoxy guanosine to dCMP, dAMP, and dGMP, respectively. Thymidine in salvage pathway has special significance to the microbial geneticist. Because the DNA is the only cellular constituent that contains thymine, it provides a route by which DNA specifically can be made radioactive.

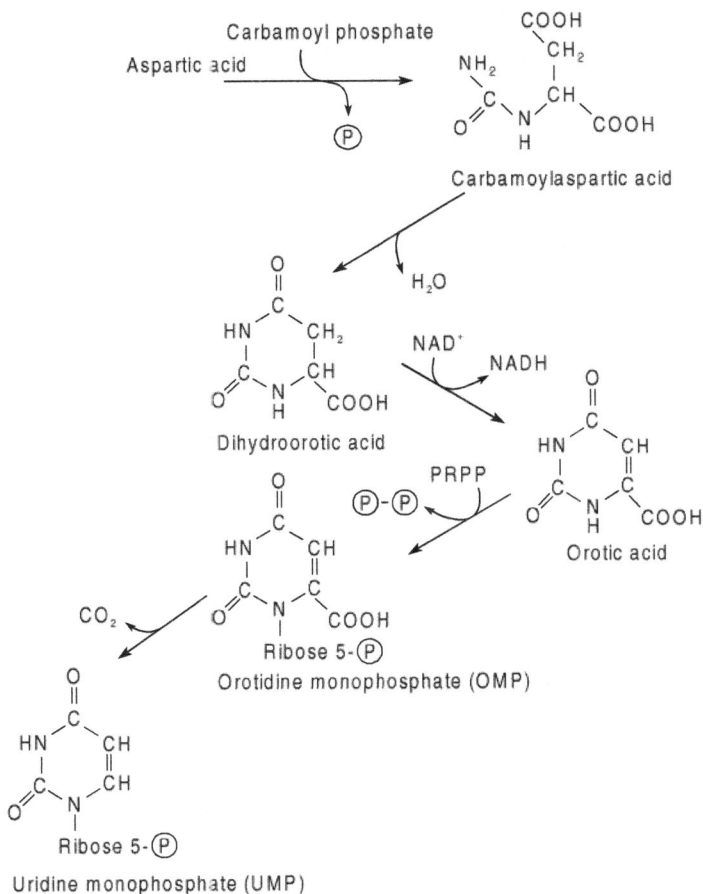

Figure 20.6   Biosynthesis of pyrimidine ribonucleotide.

In this pathway (catalysed by thymidine phosphorylase) there is an equilibrium constant between the conversion of thymine to thymidine. This will lead to the exogenous thymine in not incorporating into DNA by enteric bacteria unless steps are taken to shift the equilibrium towards the direction of biosynthesis by increasing the intracellular concentration of the second substrate, deoxyribose 1-phosphate.

This is because this compound cannot penetrate the cell membrane and the steps taken to raise its intracellular concentration must be indirect. A deoxyriboside, which does penetrate the cell and is phosphorolytically cleaved to yield deoxyribose 1-phosphate, can be added to the medium; or a genetic blockade can be introduced in the step between dUMP and dTMP. The blockage causes dUMP to accumulate, which is then degraded intracellularly to deoxyribose 1-phosphate. Many microorganisms including most pseudomonads lack completely the thymine salvage pathway.

STUDY OUTLINE

- Nucleotides are macromolecules which are essential for nucleic acid biosynthesis.
- They are the building blocks of nucleic acids both DNA or RNA.
- Nucleotides are synthesized either by de novo or salvage pathway.
- The mother molecule for de novo pathway is inosine monophosphate.
- Nucleoside is nothing but the attachment of pyrimidine bases through nitrogen atom to a pentose sugar.
- Nucleotide, can be defined as the attachment of nucleoside with a phosphate group.

CONCEPT CHECK

1. Explain the de novo pathway of nucleotide.
2. How are purines and pyrimidines synthesized?
3. Name the drugs which inhibit nucleotide biosynthesis.
4. Write about the functions of nucleotide in a system.

CRITICAL THINKING

All living forms have nucleotides for the biosynthesis of their genetic material, either DNA or RNA. But some viruses like prions are made up of only protein compounds. In this case how does the protein contribute to their survival?

21

LIPID METABOLISM

INTRODUCTION

Lipids are a group of organic molecules which are widely distributed in living systems. Lipids contribute the major part of protoplasm as they are generally hydrophobic in nature. Lipids are classified into four types.

1. Triglycerides
2. Phospholipids
3. Sterols and
4. Waxes

Triglycerides

Triglycerides consist of glycerol and fatty acid, e.g. palmitic acid, linoleinic acid. The three hydroxyl groups of the glycerol molecule is attached with three carbon chains.

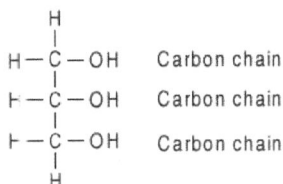

```
        H
        |
   H — C — OH    Carbon chain
        |
   H — C — OH    Carbon chain
        |
   H — C — OH    Carbon chain
        |
        H
```

If a fatty acid contains only single bonds it is called as saturated fatty acid, e.g. palmitic acid. If a fatty acid has even one double bond, it is termed as unsaturated fatty acid, e.g. linoleinic acid.

Phospholipids

These lipids contain ester group of hydroxyl alcohol (glycerol). The two hydroxyl group of glycerol is attached with carbon chain whereas, the last hydroxyl group is attached with a phosphate atom. In this structure, the phosphate head part is hydrophilic in nature, and the tail part contributes to the hydrophobic nature, which forms the lipid bilayer of the plasma membrane.

Sterols

The major sterols are alcoholic fatty acids called cholesterol which contributes to most of the organic cell wall e.g. Ergosterol, present in the cell wall of fungi.

Wax

These are esters of long chain fatty acids and long chain monohydric alcohols (or) sterols. Some organisms produce wax which contributes to their pathogenecity.

E.g. *Myco bacterium tuberculosis.* [C18:1:9 denotes a C_{18} fatty acid with a double bond between carbon 9 and 10 C18:2:9:12, denotes a C18 fatty acid with two double bonds between C_9 and C_{10} and between C_{12} and C_{13}. C18:0, denotes a C_{18} fatty acid with no double bond]. Numbering of carbon atom in fatty acid is saturated at the carboxy terminus.

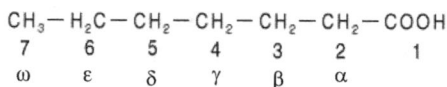

$$CH_3 - H_2C - CH_2 - CH_2 - CH_2 - CH_2 - COOH$$

7	6	5	4	3	2	1
ω	ϵ	δ	γ	β	α	

OXIDATION OF FATTY ACIDS

Fatty acids are the immediate substrates for the oxidation of fats, viz., liver, adipose tissues, muscles, heart, kidney, brain, lung, and testes. In case the fatty acids are esterified, they are first hydrolysed by esterate present in all the tissues of the body. Fatty acids are oxidized to CO_2 and water with the liberation of huge amount of energy. Several theories have been proposed to explain the mechanism of oxidation of the fatty acid chain, out of which the most important is the β oxidation theory.

β Oxidation of Fatty Acids

The principle of β oxidation of fatty acid was discovered by the German chemist, Franz. The oxidation takes place at the β carbon atom to the keto group followed by elimination of the two terminal carbon atoms

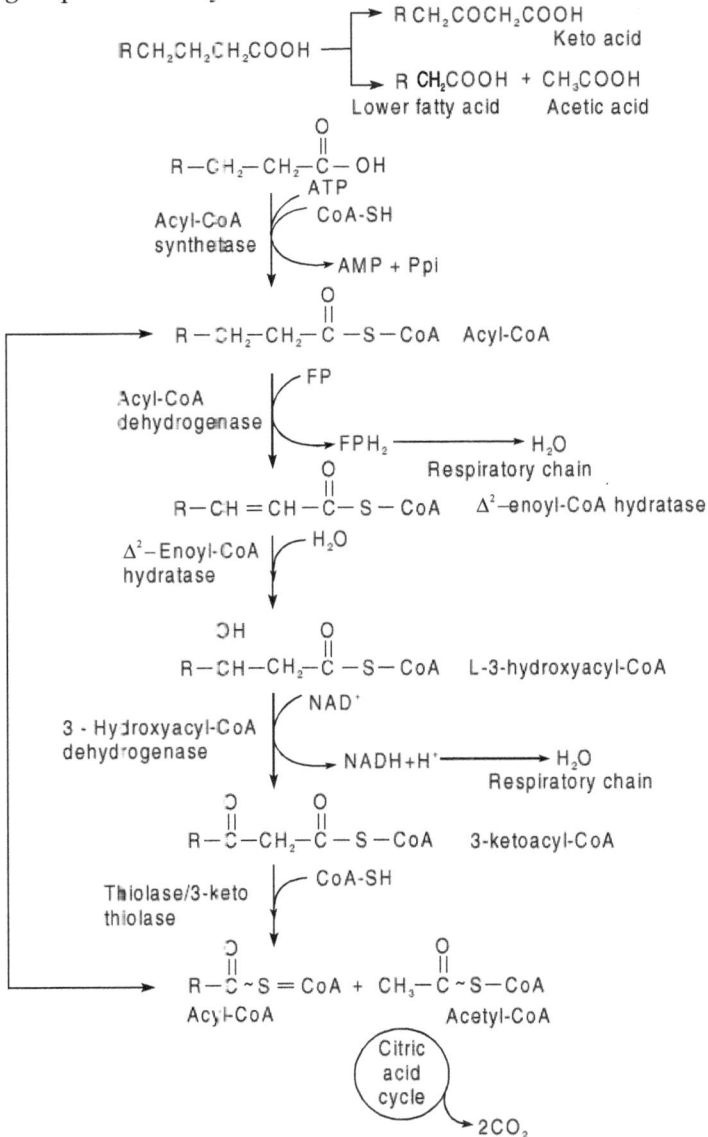

$$RCH_2CH_2CH_2COOH \longrightarrow \begin{cases} R\,CH_2COCH_2COOH \quad \text{Keto acid} \\ R\,CH_2COOH + CH_3COOH \\ \text{Lower fatty acid} \quad \text{Acetic acid} \end{cases}$$

$$R-CH_2-CH_2-\overset{\overset{\displaystyle O}{\|}}{C}-OH$$

Acyl-CoA
synthetase
ATP
CoA-SH
AMP + Ppi

$$R-CH_2-CH_2-\overset{\overset{\displaystyle O}{\|}}{C}-S-CoA \quad \text{Acyl-CoA}$$

Acyl-CoA
dehydrogenase
FP
FPH$_2$ ⟶ H$_2$O
Respiratory chain

$$R-CH=CH-\overset{\overset{\displaystyle O}{\|}}{C}-S-CoA \quad \Delta^2\text{-enoyl-CoA hydratase}$$

Δ^2-Enoyl-CoA
hydratase
H$_2$O

$$R-\overset{\overset{\displaystyle OH}{|}}{CH}-CH_2-\overset{\overset{\displaystyle O}{\|}}{C}-S-CoA \quad \text{L-3-hydroxyacyl-CoA}$$

3 - Hydroxyacyl-CoA
dehydrogenase
NAD$^+$
NADH+H$^+$ ⟶ H$_2$O
Respiratory chain

$$R-\overset{\overset{\displaystyle O}{\|}}{C}-CH_2-\overset{\overset{\displaystyle O}{\|}}{C}-S-CoA \quad \text{3-ketoacyl-CoA}$$

Thiolase/3-keto
thiolase
CoA-SH

$$R-\overset{\overset{\displaystyle O}{\|}}{C}\sim S=CoA + CH_3-\overset{\overset{\displaystyle O}{\|}}{C}\sim S-CoA$$
Acyl-CoA \qquad\qquad Acetyl-CoA

Citric
acid
cycle
⟶ 2CO$_2$

Figure 21.1 Pathway of β oxidation of fatty acid.

as acetic acid and the development of the new carboxyl group at the site of the fatty acids has two carbon atoms less than the original.

Energetics

Fatty acid is completely degraded into (C_2 units), thus palmitic acid ($C_{15} H_{31}$ COOH) is degraded to 8 acyl units in seven rounds.

$$
\begin{array}{rcl}
\text{Now since in one round one mole} & & \\
\text{of FADH}_2 & = & 2 \quad \text{moles of ATP} \\
\text{One mole of NADH+H}^+ & = & 3 \quad \text{moles of ATP} \\
\text{Total (In 1 round)} & = & \overline{5} \quad \text{moles of ATP} \\
\text{For 7 rounds total moles of ATP} = 7 \times 5 & = & 35 \\
\text{1 acetyl CoA in TCA cycle} & = & 12 \\
\text{8 acetyl CoA in TCA cycle} & = & 96 \\
\text{Total} & = & \overline{131} \\
\text{Number of ATP used for the initial} & & \\
\text{activation of fatty acid} & = & -1 \\
\text{Net ATP} & = & \overline{130}
\end{array}
$$

If one palmitic acid undergoes β oxidation it gives 130 ATP molecules.

α Oxidation

It is the removal of one carbon at a time from the carboxyl end of the molecules. It does not require CoA intermediate and does not generate high-energy phosphate.

ω Oxidation

It is a very minor pathway. It involves the hydroxylation of the CH_3 group to CH_2OH group which is ultimately oxidized to COOH group thus forming a dicarboxylic acid. Finally it forms acetyl and acyl CoA.

SYNTHESIS OF LIPIDS

Lipid synthesis is a critical process for microbes. Lipids are a class of cell constituents defined on the basis of solubility properties instead of their chemical composition. Lipids normally constitute only a small portion of the cell's weight, they are universal components of the membrane and membranous organelles of the cell. The lipid component in the prokaryotes participate in a diversity of physiological processes. Lipids are insoluble in water and soluble in nonpolar solvents such as ether, chloroform and benzene. The centrality of lipid in membrane function in a cell requires that we understand the nature of the major lipid component of the cell and the manner of their synthesis. Significant functional differences are found in the lipid found in prokaryotic and eukaryotic microbes. The presence of sterols in the membrane of eukaryotes, and with the exception of mycoplasma, the absence of these materials in prokaryotes is the basis for the selective mode of action of many membrane-directed antimicrobial chemicals. Lipids are chemically heterogeneous and include fats, phospholipids, steroids, isoprenoids and poly-β-hydroxybutyrate. However, they can be grouped into two broad classes: those that contain esterified fatty acid and those consist of repeating C5 units with structure of isoprene.

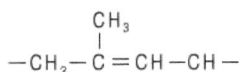

$$-CH_2-\overset{\overset{\displaystyle CH_3}{|}}{C}=CH-CH-$$

The phospho "lipids are universal membrane components of bacteria and eukaryotes. The chemical nature of the residue (x) attached to the phosphate group defines the class of phospholipid. In E. coli and Salmonella typhimurium, in which the most careful measurements have

X	Name of phospholipid
$CH_2-CH_2-NH_2$	Phosphatidylethanolamine
CH_2-CNH_2-COOH	Phosphatidylserine
$CH_2-CHOH-CH_2OH$	Phosphatidylglycerol

$$CH_2-CHOH-CH_2-O-\overset{\overset{\displaystyle O}{\|}}{\underset{\underset{\displaystyle O}{\|}}{P}}-OCH_2 \quad \overset{\displaystyle O}{\underset{}{}}$$

Cardiolipin

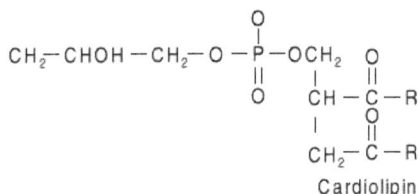

General structure of phosphoLipid

been made, the major phospholipid of the membrane is phosphatidyl ethanolamine (75%).

SYNTHESIS OF FATTY ACIDS

Fatty acids are synthesized separately and then esterified to form complex lipids. Fatty acids may be different in some bacteria. They may or may not contain straight or branched chain, double bond, OH groups or cyclopropane rings. The number of fatty acid is limited in any particular species. For example *E. coli* have six while *Bacillus subtilis* contains eight; two fatty acids are common to both species among their number. The polyunsaturated type (more than one double bond) found in eukaryotes are rare in prokaryotes. In saturated fatty acid synthesis, a protein named as acyl carrier protein (ACP) plays an important role. It is a smaller protein having a molecular weight of 10,000 which is functionally and chemically analogous to CoA. In *Clostridium* the saturated fatty acids are synthesized by the help of CoA, which carries the acyl group.

But in most cases, the acyl carrier protein carries the acyl group. The formation of long chain fatty acids starts with the transfer of an acetyl group from CoA to ACP. This complex serves as an electron acceptor to which successive C_2 units are transferred. The C_2 donor in malonyl (-) ACP is formed in turn, by decarboxylation of acetyl CoA; during transfer of the C_2 unit, CO_2 is released and free ACP is generated. The product of C_2 transfer carries a terminal acetyl group which in subsequent reactions is sequentially reduced, dehydrated and reduced again yielding an unsaturated acyl-ACP complex with two additional carbon atoms. Repetitions of this set of reactions progressively lengthen the fatty acid chain until the length characteristic of the particular bacterium. Monosaturated fatty acids are formed in various bacteria by one of the two different pathways, the aerobic pathway and the anaerobic pathway (which occur in aerobes as well as anaerobes). The aerobic pathway involves the subsequent modification of fully synthesized (but still ACP-bound) saturated fatty acids, while in the anaerobic pathway unsaturation takes place during elongation of the fatty acid chain. The aerobic pathway require the direct intervention of molecular oxygen.

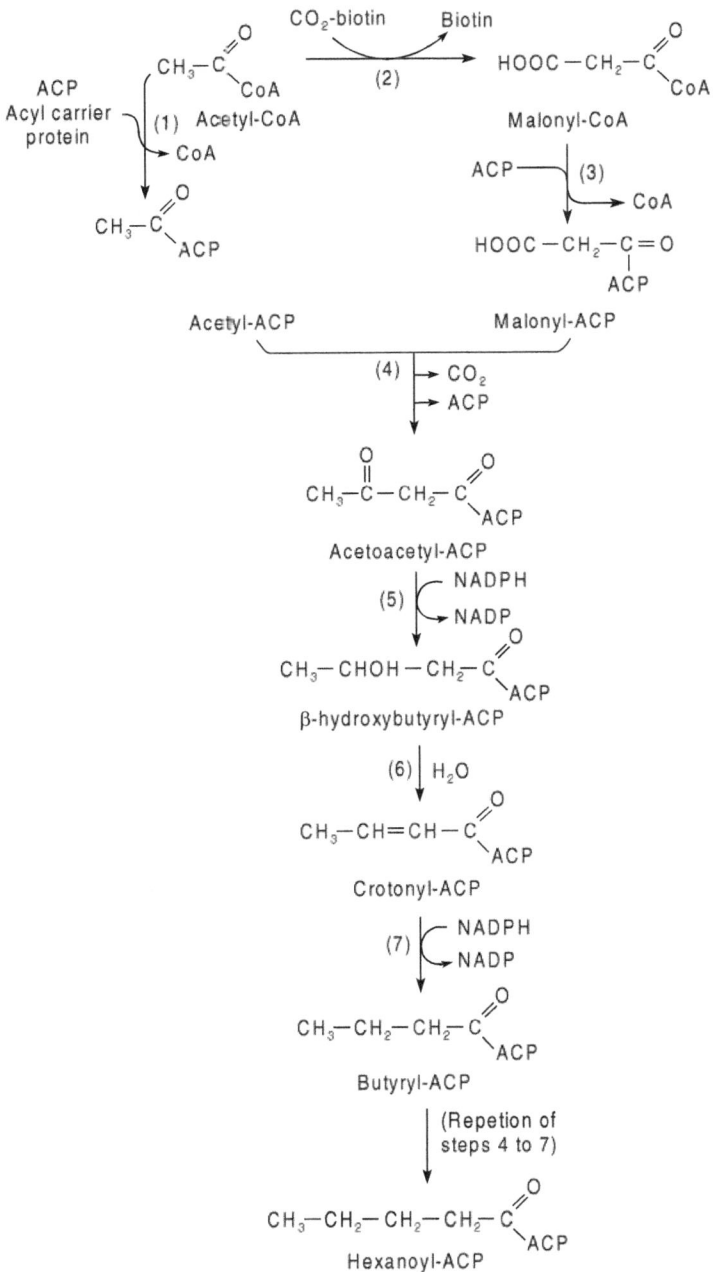

Figure 21.2 Mechanism of synthesis of saturated fatty acids.

In aerobic pathway the C_{10} hydroxyl acyl intermediate, β-OH-decanoyl-ACP can undergo normal β desaturation, leading to the formation of longer saturated fatty acid chain, or it can undergo a β-γ dehydration, leading to the homologous monosaturated fatty acids. In anaerobic pathway, the position of the double bond in the carbon chain of the eventual end product is determined by the point in biosynthesis to where it is introduced. Subsequent chain elongation leads to its location between carbon atoms 9 and 10 in C_{16} product (palmitoleic acid).

Hence, bacteria that employ the anaerobic pathway contain Cis vaccenic acid as their monosaturated C_{18} fatty acid, rather than oleic acid, the product of direct desaturation of stearic acid by the aerobic pathway.

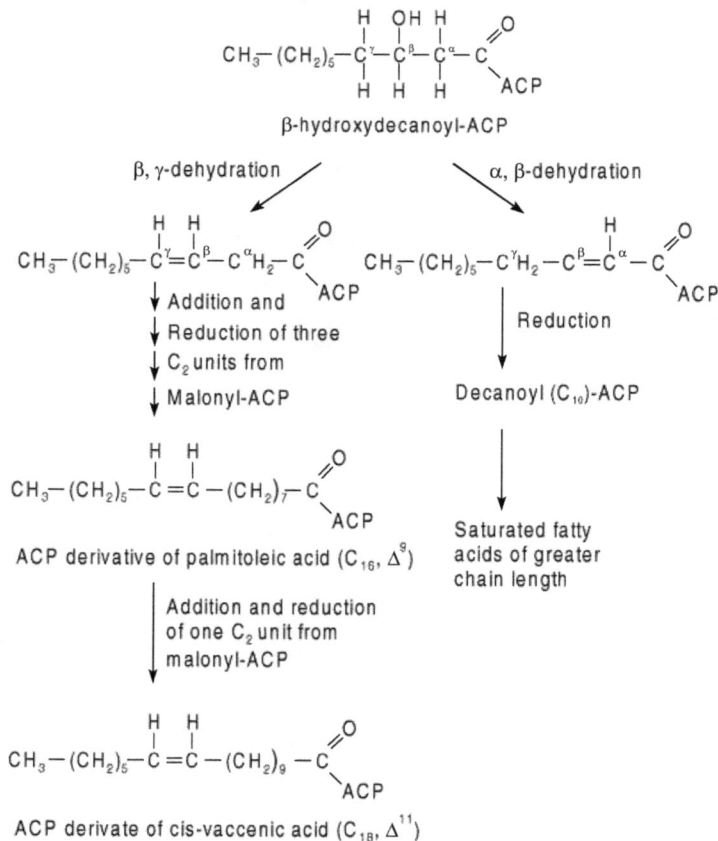

β-hydroxydecanoyl-ACP

β, γ-dehydration   α, β-dehydration

Addition and
Reduction of three
C_2 units from
Malonyl-ACP

Reduction

Decanoyl (C_{10})-ACP

ACP derivative of palmitoleic acid (C_{16}, Δ^9)

Saturated fatty acids of greater chain length

Addition and reduction of one C_2 unit from malonyl-ACP

ACP derivate of cis-vaccenic acid (C_{18}, Δ^{11})

Figure 21.3 The anaerobic pathway to monosaturated fatty acids.

SYNTHESIS OF PHOSPHOLIPIDS

Phospholipids are synthesized from fatty acids and precursors. Trisose-
phosphate is reduced by the dihydroxyacetone phosphate to

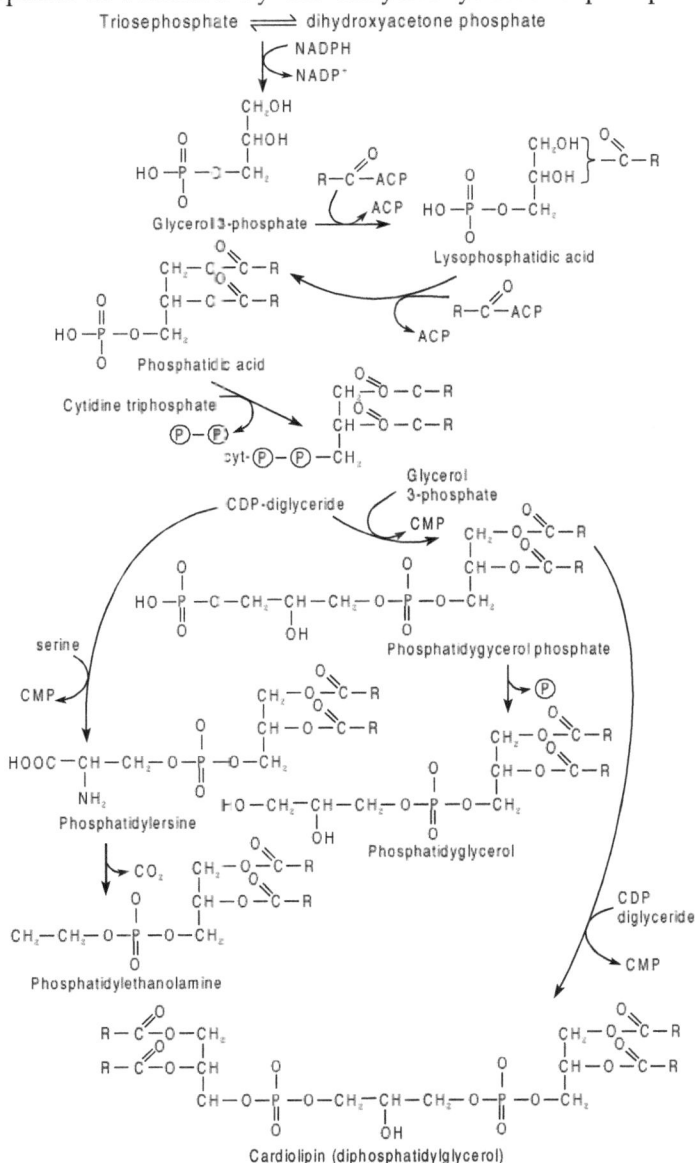

Figure 21.4 Pathway of formation of the major phospholipid classes found in *E. coli*.

3-glycerophosphate which is subsequently esterified by two fatty acid residues. The resulting diglyceride, phosphatidic acid is then activated by CTP to form CDP-diglyceride, which undergoes transfer reactions with serine and a glycerophosphate, releasing CMP. The decarboxylation product of the phospholipid class is the phosphatidylethanolamine. The reaction between CDP diglyceride and α-glycerol phosphate leads to the other phospholipid classes, phosphatidyl glycerol and cardiolipin.

STUDY OUTLINE

• Lipids are a class of cell constituents defined on the basis of solubility properties instead of their chemical composition. They constitute only a small portion of the cell weight. They are universal components of the membrane and membranous organelles of the cell.

• Significant differences are found in the lipids of prokaryotes and eukaryotes.

• Phospholipids are the universal membrane components of bacteria and eukaryotes.

• Lipids contribute to the major part of the protoplasm.

• Lipids are classified into 4 types. They are triglycerides, phospholipids, sterols and wax.

CONCEPT CHECK

1. Explain the biosynthesis of phospholipids.
2. How are lipids oxidized through β-oxidation?
3. Write about the α and ω oxidation of lipids.
4. Explain the mechanism of biosynthesis of saturated fatty acids.

CRITICAL THINKING

In *Mycobacterium tuberculosis*, the cell wall is made up of polyhydroxybutyric acid, a special type of lipid soluble in ethanol or acetone. How can you stain this group of organisms?

22

ANAEROBIC RESPIRATION

INTRODUCTION

The anaerobic energy yielding-process in which the electron transport chain acceptor is an oxidized organic molecule other than O_2 is called anaerobic respiration. The major electron acceptors are nitrate, sulphate and CO_2. These electrons are derived from organic molecules and are usually donated either to organic electron acceptor through fermentation or utilized by electron transport chain (aerobic respiration). Some bacteria have electron transport chains that can operate with inorganic electron transport acceptors other than O_2.

DENITRIFICATION

Some bacteria can use nitrate as the electron acceptor at the end of their electron transport chain and still produce ATP. Nitrate may be reduced to nitrite by nitrate reductase, which replaces cytochrome oxidase.

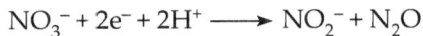

$$NO_3^- + 2e^- + 2H^+ \longrightarrow NO_2^- + N_2O$$

However, reduction of nitrate to nitrite is not a particularly effective way of making ATP, because a large amount of nitrate is required for growth (a nitrate molecule will accept only two electrons). The nitrate formed is also quite toxic, therefore, nitrate often is further reduced all the way to nitrogen gas, a process known as denitrification. Each nitrate will then accept five electrons, and the product will be nontoxic.

$$2NO_3^- + 12e^- + 12H^+ \longrightarrow N_2 + 6H_2O$$

Denitrification is carried out by some members of the genera *Pseudomonas* and *Bacillus*. They use this route as an alternative to normal aerobic respiration and may be considered as facultative anaerobes. If O_2 is present, these bacteria use aerobic respiration (the synthesis of nitrate reductase is repressed by O_2). Denitrification in anaerobic soil results in the loss of soil nitrogen and has adverse effects on soil fertility.

ANAEROBIC RESPIRATION IN METHANOGENS

Two major groups of bacteria employ anaerobic respiration. They are termed as obligate anaerobes. These organisms use CO_2 or carbonate as a terminal acceptor and are called as methanogens because they reduce CO_2 to methane. Sulphate also can act as the final acceptor in bacteria, for example, *Desulfovibrio* which reduces sulphate to sulphide (S^{2-} or H_2S) and 8 electrons are accepted.

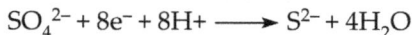

$$SO_4^{2-} + 8e^- + 8H+ \longrightarrow S^{2-} + 4H_2O$$

ENERGY PRODUCTION

Anaerobic respiration is not as frequent an event in ATP synthesis as aerobic respiration, i.e., not much ATP is produced by oxidative phosphorylation with nitrate, sulphate, or CO_2 as the terminal acceptor. Reduction in ATP yield, arises from the fact that the alternative electron acceptors like nitrate, sulphate or CO_2 have less positive reduction potential than O_2 molecules. The reduction potential differences between a donor like NADH and nitrate is smaller than the difference between NADH and O_2. Because energy yield is directly related to the magnitude of the reduction potential difference and so less energy is available to make ATP in anaerobic respiration. Nevertheless, anaerobic respiration is useful because it is more efficient than fermentation and allows its possessor to make ATP by electron transport and oxidative phosphorylation.

STUDY OUTLINE

- Anaerobic respiration is defined as an energy-yielding process in which the electron transport chain acceptor is an oxidized inorganic molecule other than O_2.

- Various microorganisms perform this type of respiration.

- Some bacteria involved in nitrogen fixation carry out this respiration, for example *Pseudomonas* and *Bacillus*.

- Methanogens are strict anaerobes and use CO_2 as their terminal acceptor by a process called methanogenesis.

CONCEPT CHECK

1. Define denitrification.
2. What is anaerobic respiration?
3. How do anaerobic organisms uptake energy through anaerobic respiration?

CRITICAL THINKING

Anaerobic respiration is a less energy yielding process when compared to other energy-yielding pathways. How do aerobic microorganisms satisfy all their needs?

23

TRANSPORT MECHANISM IN MICROBES

INTRODUCTION

Transport is a critical activity of microbes. This mechanism allows the organism to take up nutrients from the external environment. However, transport is also important for the removal of toxic materials from the internal environment, allowing the cell to maintain an intracellular environment compatible with life.

Transport may be distinguished into two types—nutrient transport and electron transport. These two types of transport are intertwined. However, nutrient transport concerns uptake and removal of materials associated with the formation and function of cells, and electron transport concerns the generation of energy.

TYPES OF TRANSPORT

The transport of substances may be viewed from a number of perspectives. Except simple diffusion, transport always involves proteins, but the manner in which proteins facilitate the transport process varies from system to system. Based on the direction, the protein-mediated transport occurring in a single direction is called uniport. However, a protein may mediate the transport of more than one material. When this is the case and the transport of two materials occurs in the same direction, it is referred to as symport. If the same protein mediates simultaneously the transport of two materials, but the transport of one material is in the net inward direction while the second material is in the net outward direction, the process is referred to as antiport.

Simple Diffusion

Simple diffusion is a transport mechanism where there is no utilization of energy and where protein is not mediated. The direction of transport is dictated solely by the extra- and intracellular material concentration. If the intracellular material concentration is more than that of the extracellular environment, movement proceeds in the net outward direction. Similarly if the external substance concentration exceeds its internal concentration, net transport is inward. In simple diffusion, accumulation against the concentration is impossible.

Facilitated Diffusion

This is different from simple diffusion but like simple diffusion it is also concentration-dependent. It does not require biologically derived energy, but it is protein-mediated. Its velocity is within limits, a function of the difference in concentrations between the cell exterior and interior. This is because, at low material concentration, only a portion of the potential reactive sites on carrier molecules are occupied at a given moment in time. As the concentration of transported material is increased and the proportion of total transport is increased, the proportion of total transport sites occupied increases, with a corresponding increase in transport rate. Eventually, the concentration is such that all of the available sites for transport are occupied. At this point, further increase in the concentration of the transported material does not result in an increase in transport rate. Facilitated diffusion differs from simple diffusion in requiring a protein but the net effects of facilitated and simple diffusion are the same. In both cases, net movement is from a region of higher concentration of the transported substance to a region of lower concentration and the net transport is a function of the difference in concentration.

Active Transport

Active transport is the most frequently encountered mode of transport in microbes. This system is intimately related to electron transport, which usually provides the force by which active transport is accomplished. This electron usually generates the energy for active transport. Some anaerobes generate transport energy in the absence of electron transport. Such organisms obtain energy for transport either by ATP hydrolysis or by the symportic transport of fermentation end products and protons to the cell exterior. The transport of substances by electron transport produces a proton motive force (PMF) which is composed of a proton gradient and a membrane potential. In many, if not most, systems, the bulk of the total PMF is attributed to the proton gradient or a secondary gradient generated by it. Active transport differs from simple diffusion in its need for protein-mediation, and therefore, exhibits saturation kinetics. Active transport also exists in nutrient-sparse environment. Active transport also requires biologically active energy which is not required for either simple or facilitated diffusion.

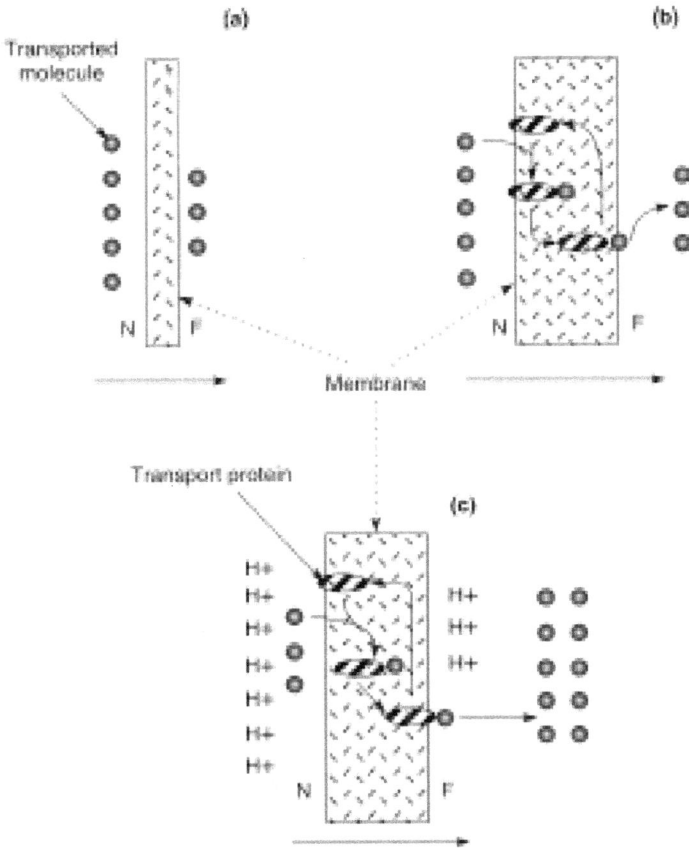

Figure 23.1 Elements of simple diffusion (a), facilitated diffusion (b) and active transport (c) The letters N and F indicate the near and far sides of the membrane, respectively. The horizontal arrows indicate the direction of net transport.

CHEMICAL MODIFICATION TRANSPORT

In active transport, simple diffusion, and facilitated diffusion, the transported material is unaltered during transport, but not in the case of chemical modification transport. Also, active transport and diffusion of any kind are primarily membrane-mediated phenomena. But in chemical transport system, both membrane and the cytoplasm are intimately involved in the transport process. This transport is commonly known as the phosphotransferase system (PTS). The chemical modification is a multicomponent system, whose components are inducible, that is formed only under certain conditions and some are constitutive, that is formed under all conditions continuously with regard to the environment. They are found exclusively in prokaryotes and have been studied most thoroughly with sugars and sugar alcohols.

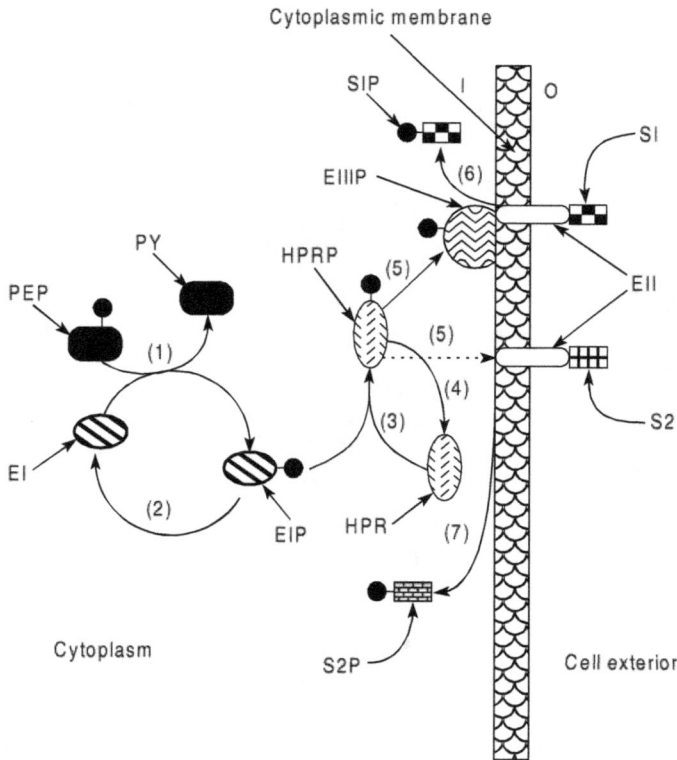

Figure 23.2   The essential elements of chemical modification transport.

Mechanism of Chemical Modification

In phosphotransferase system, initially the phosphoenol pyruvate reacts with a cytoplasmic, constitutive protein known as enzyme I (EI) to produce phosphorylated form of enzyme I (ETP). The phosphorylated enzyme I (ETP) then reacts with a second constitutive cytoplasmic protein, histidine protein (HPR), yielding histidine protein phosphate (HpRP). Next the membrane-associated protein enzyme II (EII) comes in contact with the (HpRP), producing enzyme II phosphate (EIIP). Finally, EIIP reacts within the membrane with the transported material, converting it to a phosphorylated form. In some systems, yet another protein, enzyme III (EIII) is involved in the transport process. When such is the case, HpRP transfers its phosphate to EIII, producing enzyme III phosphate to produce EIIIP, which, in turn phosphorylates the transported substance.

Regulation of phosphotransferase systems is extremely complex. The state of cellular energy metabolism exerts a substantial effect on the process since the rate and extent of formation of PEP dictates the phosphorylation state of the remaining components of the system. The nature of the external environment also regulates the system. EII and EIII are specifically induced by particular materials, and are distinguished by subscripts that denote the substance that leads to their induction. Thus EII gluc is induced by the presence of glucose and EII gal is formed in response to galactose. Because EII and EIII proteins are involved, another level of regulation is possible. Although both EII and EIII proteins are inducible, the specificity of their interaction with transport molecules is not absolute. The EII gluc protein, for example, may react with lesser affinity, with a molecule other than glucose, perhaps galactose. Regulation of the operation of the PTS system may reflect not only the presence of particular substances, but when multiple transport substrates are present, their relative concentrations as well. Competition for HprP by various EII and EIII proteins is yet another level at which control may be exerted. Finally, the intracellular concentration of phosphorylated transported materials exerts a diversity of regulatory effects on the system.

TRANSPORT IN GRAM-NEGATIVE BACTERIA

Earlier, it was believed that the transport process is entirely a membrane-mediated phenomenon. But the study of phosphotransferase system changed this idea of transport and it is now accepted that the other cellular parts are also involved in the transport processes, e. g., the cytoplasm. Because of the selective permeability of the cell membrane, little bit attention is also given to the cytoplasm. In gram-negative bacteria the periplasmic space and cell envelope are intimately involved in the transport process. The cytoplasmic membrane is bounded by a second, fundamentally hydrophobic layer, the outer membrane. Any material that would enter or leave the gram-negative bacterium must traverse, not only the cytoplasmic membrane, but also the outer membrane. The cytoplasmic and outer membranes of gram-negative bacteria contain proteins that serve as channels for transport of hydrophilic substances. Some outer membrane proteins (OMP) in the organism also facilitate the transport of hydrophilic substances and the protein is abbreviated as OMP, followed by a capital letter (e.g. OMPA). Most of the membrane protein involved in the transport system are multifunctional. For example, some of the proteins participate not only in transport but also in other processes. "OMPA" serves as receptor for phage Tula, participates in an interaction with the lipopolysaccharide of the outer membrane, and helps to stabilize mating pairs in F factor-mediated conjugation.

Another example for the multifunctional property of membrane is "OMPB" protein which serves as a channel for the diffusion of maltose and other substances. "OMPC" serves as a diffusion channel for a number of small molecules and also as a receptor for both phages Tulb and T4. "OMPF" serves as a general channel for small molecules and in addition as a receptor for phages Tula and T2. The outer membrane proteins are obligately trimers. But components of the various trimers exist in different physical relationships to each other with the membrane. Trimeric outer membrane transport proteins are collectively referred to as porins.

ION TRANSPORT

In addition to the organic compound transport, outer membrane also transports the inorganic ions like Fe^{2+} and PO_4^{3-}. This transport is mediated by both outer membrane and periplasmic proteins. Thus, the PhoE outer membrane protein facilitates anion transport when phosphate is limited. But in *E.coli* the periplasmic binding protein (PhoS) is involved in this transport. But the sphaeroplasts do not contain the periplasmic binding protein. In addition to phoS-mediated

PO_4^{3-} transport in *E.coli*, the organism also contains a phosphate transport system (PIT system) that does not require the phoS protein and operates in normal sphaeroplasts.

Transport systems involving periplasmic components can often be distinguished from those that do not by subjecting the suspected organism to osmotic shock, which releases periplasmic components. If transport is abolished or altered by osmotic shock treatment, one may suspect that periplasmic components are involved in the transport process. If osmotic shock treatment does not alter transport, it is likely that the periplasm is not involved in the system under study.

IRON TRANSPORT

Iron is an essential and critical substance for all life, and is very much needed in energy metabolism. The competition between organisms for iron is a mode of pathogenesis for some organisms. For iron transport, specific organisms have specific iron-transport proteins commonly known as siderophores. There are a great number of siderophores which include, hydroxamates or catechol compounds.

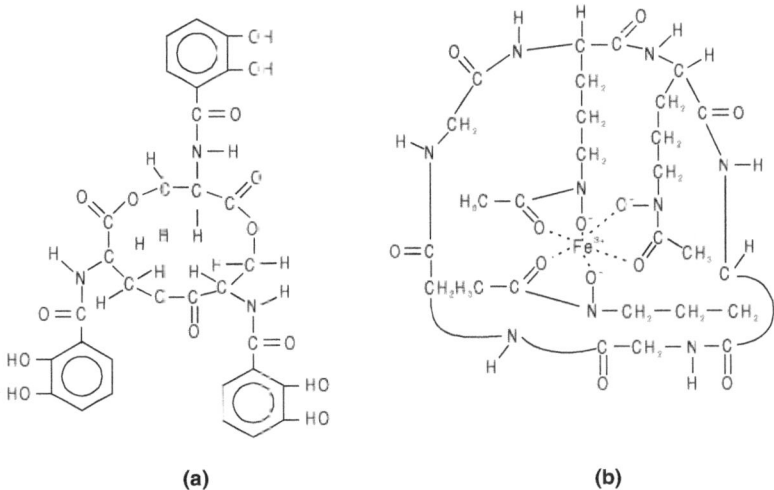

(a)   (b)

Figure 23.3 The general structure of (a) a catechol (enterochelin) and (b) a hydroxamate siderophore (ferrichrome).

Both siderophores bind the Fe^{3+} ion by chelation, but the precise mechanism by which the bound iron enters the cell varies among organisms. Certain organisms transport iron without entry of the siderophore into the cell by incorporating cell-associated reduction of organically bound iron. In other situations, for example in *Bacillus megaterium*, the intact ligand-bound iron complex may enter the cell by facilitated diffusion. Finally, in some organisms, more than one mechanism for iron uptake is known.

STUDY OUTLINE

- Transport is the mechanism which aids microorganisms in the uptake of nutrients.

- There are two types of transport—nutrient transport and electron transport.

- Uniport refers to the protein-mediated transport occurring in a single direction. If the same protein mediates simultaneously the transport of two materials, it is termed as symport.

- Materials can be transported either by simple diffusion or facilitated diffusion.

- In active transport, proton motive force is generated and ATP is synthesized.

- In active transport, chemical modification is induced. In the case of prokaryotes, the gram-positive chemical modification is different from that of gram-negative. This is because of the constituents of cell membrane and structure of cell wall.

CONCEPT CHECK

1. Define uniport, symport and antiport.
2. Explain the processes of simple and facilitated diffusion.
3. Explain the mechanism of active transport.
4. Explain the chemical modification process.
5. How do gram-positive and gram-negative bacteria differ in their active transport?

CRITICAL THINKING

Microorganisms transport their nutrients through cell membrane. In transport mechanism proteins, sugars and amino acids are transported. Iron transporting microorganisms chelate the iron compounds. However some microorganisms acumulate heavy metals like gold, chromium and mercury in their internal structure as "nanoparticles". How do microorganisms mediate this type of transport?

24

NITROGEN CYCLE

Microorganisms present in the soil serve as biogeochemical agents which convert the complex organic compounds into simple inorganic compounds or into their constituents elements. This overall process is called mineralization. This conversion gives inorganic compounds which serve as nutrients for plants and animals including man. For many microorganisms, growth and reproduction are limited by the availability of utilizable nitrogen because nitrogen is one of the major elements for microorganisms. They act as building blocks for nucleotides and amino acids. Many microorganisms cannot utilize inorganic nitrogen compounds. But all microorganisms can convert ammonia (NH_3) to organic nitrogen compounds, i.e., substances containing C–N bonds. However, not all microorganisms utilize inorganic form of nitrogen like nitrogen gas (N_2), the most abundant form of nitrogen in the atmosphere. The nitrate ion (NO_3^-), the product of nitrogen fixation, is a soil constituent essential for the growth of most plants. The reduction of N_2 to NH_3 is called biological nitrogen fixation. This reduction is carried out only by certain microorganisms, but the reduction of NO_3 to NH_3 by contrast, is widespread among both plants and microorganisms.

As in the consideration of any limited resource, it is useful to think about nitrogen metabolism in terms of an economy, a nitrogen economy that focuses on question of supply, demand, turnover, reuse, growth, and maintenance of a steady state. Within the biosphere, a balance is maintained between total inorganic and total organic forms of nitrogen. The conversion of inorganic to organic nitrogen starts with nitrogen fixation and nitrate reduction.

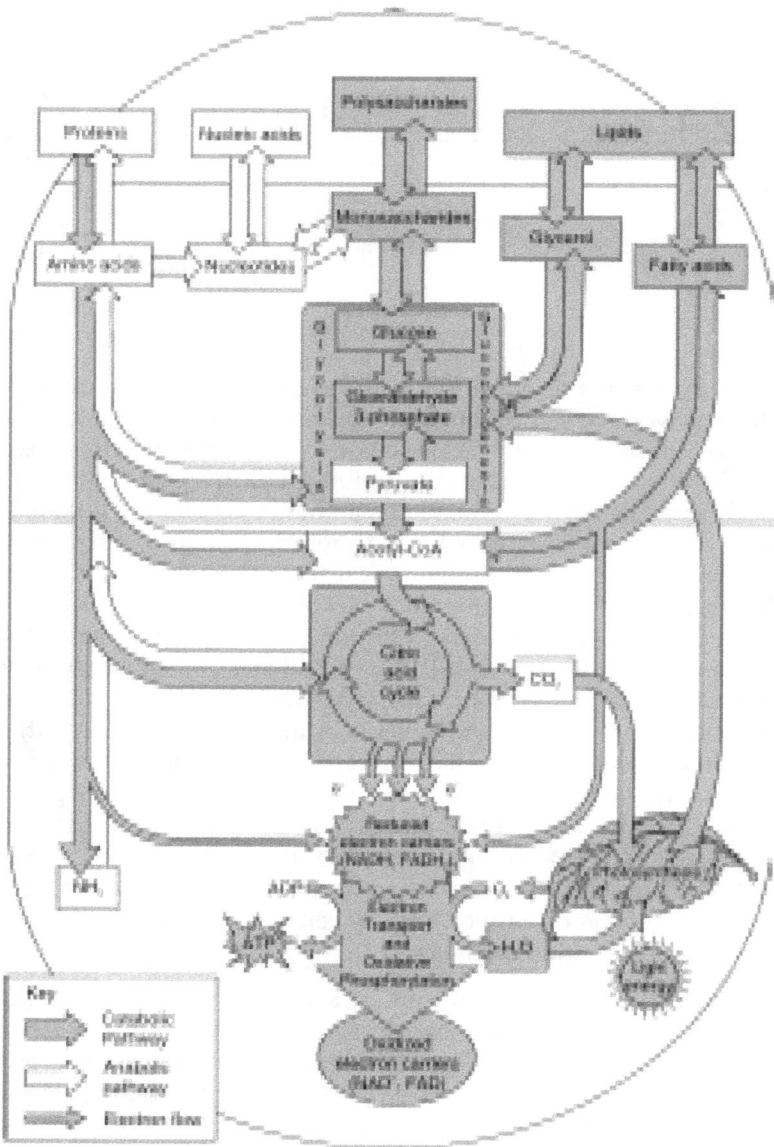

Figure 24.1   Pathways of nitrogen metabolism.

The sequence of changes from free atmospheric nitrogen to fixed inorganic nitrogen, to simple organic compounds to complex organic compounds in the tissues of plants and animals, and the eventual release of this nitrogen back to the atmosphere is termed as "nitrogen cycle".

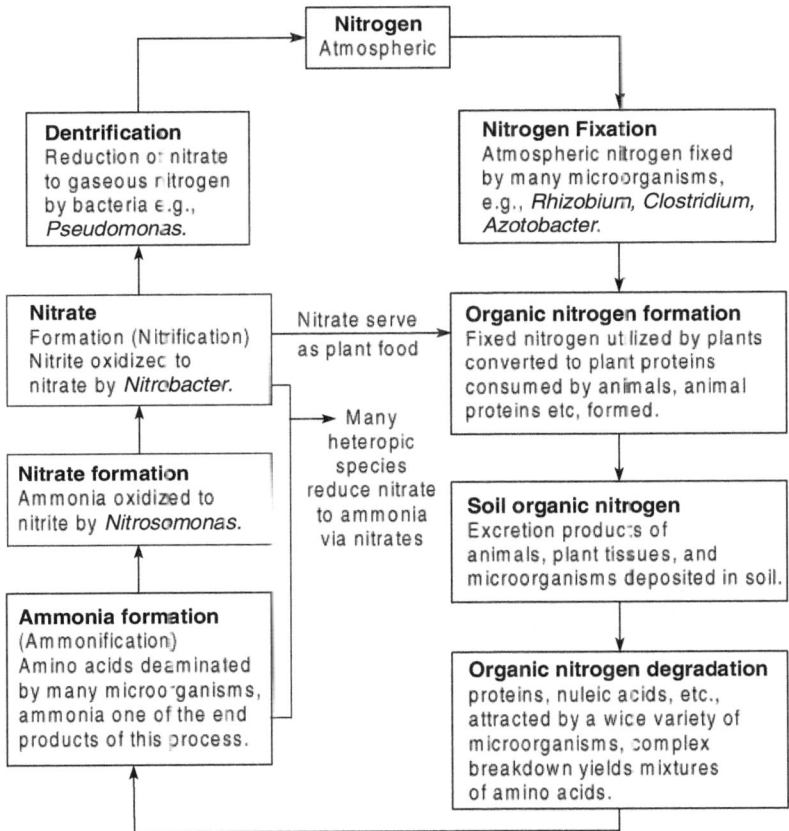

```
                          ┌─────────────────┐
              ┌──────────►│    Nitrogen     │◄──────────┐
              │           │  Atmospheric    │           │
              │           └─────────────────┘           │
              │                                          │
    ┌─────────────────────┐              ┌─────────────────────────────┐
    │  Dentrification     │              │  Nitrogen Fixation          │
    │  Reduction of nitrate│             │  Atmospheric nitrogen fixed │
    │  to gaseous ritrogen │             │  by many microorganisms,    │
    │  by bacteria e.g.,   │             │  e.g., Rhizobium, Clostridium,│
    │  Pseudomonas.        │             │  Azotobacter.               │
    └─────────────────────┘              └─────────────────────────────┘
              ▲                                          │
    ┌─────────────────────┐  Nitrate serve  ┌─────────────────────────────┐
    │  Nitrate            │  as plant food  │  Organic nitrogen formation │
    │  Formation (Nitrification)           │  Fixed nitrogen utilized by plants│
    │  Nitrite oxidizec to                 │  converted to plant proteins│
    │  nitrate by Nitrobacter.│  Many       │  consumed by animals, animal│
    └─────────────────────┘  heteropic     │  proteins etc, formed.      │
              ▲              species        └─────────────────────────────┘
    ┌─────────────────────┐  reduce nitrate                │
    │  Nitrate formation  │  to ammonia     ┌─────────────────────────────┐
    │  Ammonia oxidized to │  via nitrates  │  Soil organic nitrogen      │
    │  nitrite by Nitrosomonas.│            │  Excretion products of      │
    └─────────────────────┘                 │  animals, plant tissues, and│
              ▲                              │  microorganisms deposited in soil.│
    ┌─────────────────────┐                 └─────────────────────────────┘
    │  Ammonia formation  │                                │
    │  (Ammonification)    │                ┌─────────────────────────────┐
    │  Amino acids deaminated│              │  Organic nitrogen degradation│
    │  by many microorganisms,│             │  proteins, nuleic acids, etc.,│
    │  ammonia one of the end │             │  attracted by a wice variety of│
    │  products of this process.│           │  microorganisms, complex    │
    └─────────────────────┘                 │  breakdown yields mixtures  │
              ▲                              │  of amino acids.            │
              └──────────────────────────── └─────────────────────────────┘
```

Figure 24.2 Nitrogen cycle in nature.

The biomolecules like purines, nucleic acids, proteins, and pyrimidines are deposited in the form of animal and plants wastes or other tissues and are deposited in the soil as nitrogenous substances.

We will discuss some of the biological reactions in the nitrogen cycle.

The first step in nitrogen cycle is proteolysis. Since nitrogen compounds are locked in the form of complex molecules, the first step is to unlock the nitrogen in order to set this organically bound nitrogen free for reuse. Enzymatic hydrolysis of proteins (proteolysis) is accomplished by microorganisms, which have the capacity to elaborate extracellular proteinase that converts the protein to smaller units (peptides). The peptides are then attacked by peptidases, resulting ultimately in the release of individual amino acids. The complete reaction may be summarized as below:

$$Protein \xrightarrow{Proteinase} Peptides \xrightarrow{Peptidases} amino\ acids$$

Some of the bacterial species like clostridia, e.g., *Clostridium histolyticum* and *C. sporogenes* have large amounts of active enzyme for protein hydrolysis. Some species like *Proteus, Pseudomonas* and *Bacillus* have a less degree of activity. Many fungi and soil actinomycetes are extremely proteolytic. Peptidases, however, occur widely in microorganisms as demonstrated by the fact that peptones (partially hydrolysed proteins) are a common constituent of bacteriological media and provide a readily available source of nitrogen.

The end products of proteolysis are amino acids and these amino acids are subjected to a variety of pathways for microbial decomposition, with the liberation of nitrogen from these compounds and this is accomplished by the deamination reactions, i.e., removal of the amino group. Although several variations of deamination reactions are exhibited by microorganisms, one of the end product is always ammonia, NH_3. An example of a specific deamination reaction is

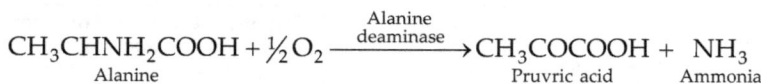

$$CH_3CHNH_2COOH + \tfrac{1}{2}O_2 \xrightarrow[deaminase]{Alanine} CH_3COCOOH + NH_3$$
$$\quad\text{Alanine} \qquad\qquad\qquad\qquad\quad \text{Pruvric acid} \quad \text{Ammonia}$$

The production of ammonia from the amino acids by oxidative deamination is called as ammonification. The variation of ammonia production depends upon the soil condition. Because, ammonia is highly volatile, when released in the soil, it easily solubilizes.

The next step is the conversion of ammonia to nitrate by microorganisms, and the process is called "nitrification" and occurs in two steps, each step being performed by different groups of bacteria.

1. Oxidation of ammonia to nitrite by ammonia-oxidizing bacteria

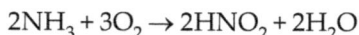

$$2NH_3 + 3O_2 \rightarrow 2HNO_2 + 2H_2O$$

2. Oxidation of nitrite to nitrate by nitrite-oxidizing bacteria

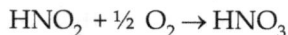

$$HNO_2 + \tfrac{1}{2}\ O_2 \rightarrow HNO_3$$

The group of ammonia-oxidizing organisms and nitrite-oxidizing organisms mostly belong to gram-negative chemolithotrophs. Chemolithotrophs derive their source of carbon through carbon dioxide fixation and energy is derived by the oxidation of NH_3 or NO_2^- depending upon the group. The nitrifying bacteria occur widely in nature in a variety of habitats including soil, sewage, and aquatic environments.

Nitrifying bacteria cannot be isolated easily like other heterotrophs because they may be present in very small numbers compared to heterotrophs and other physiological types. Some enrichment cultures are used for their isolation. Large amounts of inoculum are needed for the isolation. They can be incubated in the dark at 25 to 30°C for a period of 1 to 4 months. Ammonia-oxidizing bacterial species vary in morphology (rod, spherical, spiral or lobular) and usually have an extensive membrane system within the cytoplasm. They form cysts and zooglea.

Nitrosomanas luropaea and *Nitrosococcus oceanus* are the examples for ammonia oxidizers.

Some heterotrophic bacteria are capable of converting nitrates into nitrites or ammonia. This process normally occurs in anaerobic conditions, e.g. in waterlogged soil. The oxygen of the nitrate serves as an acceptor for electrons and hydrogen. The process involves several reactions, and the overall result is

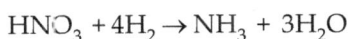

$$HNO_3 + 4H_2 \rightarrow NH_3 + 3H_2O$$

The next step is the transformation of nitrogen to gaseous nitrogen and is accomplished by microorganisms in a series of biochemical reactions. This process is called denitrification. Some genera of bacteria are capable of transforming NO_3^- to N_2 e.g., *Achromobacter, Agrobacterium, Bacillus, Flavobacterium*.

The overall biochemical reaction which expresses the process of denitrification is

$$\underset{\text{Nitrate}}{2NO_3^-} \rightarrow \underset{\text{Nitrite}}{2NO_2^-} \rightarrow \underset{\text{Nitric oxide}}{2NO} \rightarrow \underset{\text{Nitrous oxide}}{N_2O} \rightarrow \underset{\text{Nitrogen}}{N_2}$$

Environmental conditions in soil have a significant effect on the level of denitrification. For example, the process is enhanced in soils (i) with an adundance of organic matter, (ii) with elevated temperature (25 to 60°C) (iii) with neutral or alkaline pH. Oxygen is necessary for nitrite and nitrate formation.

STUDY ONLINE

- Nitrogen metabolism plays an important role in plants, animals and microorganisms. Nitrogen is the major source for amino acids, nucleic acids, and pyrimidine and purine synthesis. Nitrogen metabolism includes the fixation of molecular nitrogen in the soil.

- The cycle starts with proteolysis and ends with denitrification.

- In denitrification, the produced nitrate is again converted to molecular nitrogen (N_2).

- For each step in the process, there are various anaerobic and aerobic microorganisms that are involved.

CONCEPT CHECK

1. Define proteolysis and enzymes involved in proteolysis.

2. What do you mean by denitrification?

3. How do ammonification and nitrification play a major role in nitrogen fixation?

4. Discuss in detail about the nitrogen cycle.

CRITICAL THINKING

1. Nitrogen cycle shows that various microorganisms are involved. How can these varied microorganisms grow in the soil, when there would be some competition within these diverse groups.

2. Each and every step in the nitrogen cycle involves the oxidation process. Is the energy produced from this cycle higher, compared to other reactions?

25

ASSIMILATION OF NITROGEN AND SULPHUR

There are six major bioelements in the environment—carbon, nitrogen, sulphur, hydrogen, phosphorus and oxygen. But out of these six elements, nitrogen and sulphur lack precursor metabolites. These elements are needed for the cellular constituents. So they are incorporated into the cellular constituents as a consequence of certain reactions in the biosynthetic pathways. These elements enter into the metabolic pathways in the reduced state—nitrogen as ammonia (NH_3) and sulphur as hydrogen sulphide (H_2S). But they are found in the environment only in the oxidized state.

AMMONIA ASSIMILATION PROCESS

Ammonia is found in the organic compounds and in inorganic form at different oxidation states. Every metabolic reaction like nucleotide biosynthesis and amino acid biosynthesis needs nitrogen source. But it can be satisfied by fixation reaction. Nitrogen atom in ammonia has the same oxidation level as the nitrogen in the organic constituents of the cell. In the case of ammonia assimilation, oxidation state of nitrogen atom is not possible. For that the fixation of ammonia is needed for the cell metabolism. It can be done through three fixation reactions.

The first reaction involves formation of an amino group in glutamic acid.

The other two reactions involve the formation of amido groups in asparagine and glutamine.

i.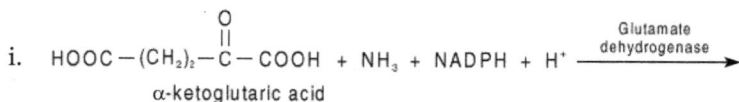
α-ketoglutaric acid

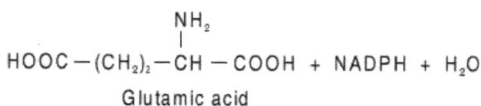
Glutamic acid

ii.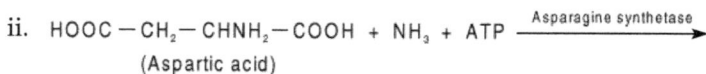
(Aspartic acid)

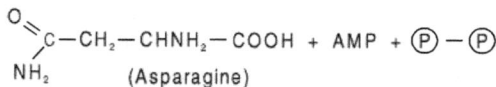

iii.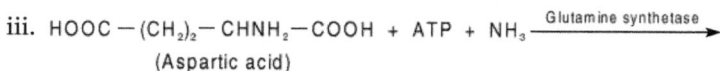
(Aspartic acid)

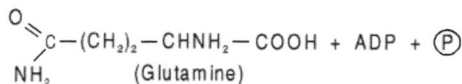

The three amino acids involved in this process (asparagine, glutamic acid and glutamine) are the direct precursors of proteins. But the amino acid asparagine serves in this fixation. However, both glumatic acid and glutamine play additional role for the transfer of amino and amido groups to all other nitrogenous precursors of cellular macromolecules. For example, the amino acids alanine, aspartic acid, and phenylalanine are formed by transamination between glutamic acid and non-nitrogenous metabolites.

i.e.

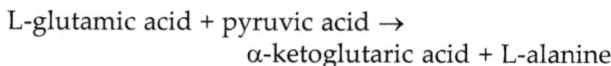

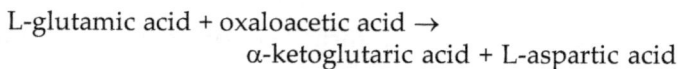

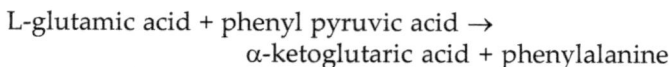

The amino group of glutamine is the source of the amino groups of cytidine triphosphate, carbamoyl phosphate, NAD, and guanosine triphosphate, among others; e.g.,

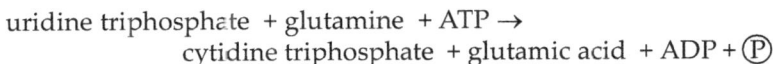

uridine triphosphate + glutamine + ATP →
    cytidine triphosphate + glutamic acid + ADP + $\textcircled{P}$

In a cell, the syntheses of glutamic acid and glutamine depend on the concentration of NH_3 and availability in the cell. If the concentration is high, two enzymes like dehydrogenase and glutamine synthetase catalyse two essential sequence reactions which lead to the synthesis of the following two compounds,

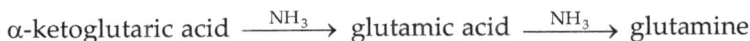

α-ketoglutaric acid $\xrightarrow{\text{NH}_3}$ glutamic acid $\xrightarrow{\text{NH}_3}$ glutamine

In this reaction ketoglutarate dehydrogenase has low affinity to the substrate NH_3 (ammonia). At that low concentration of NH_3, the enzyme ceases to function and thus the reaction becomes inoperative. In this situation, the cell synthesizes a new enzyme called glutamate synthase, sometimes called as GOGAT (an acronym for the alternative name glutamine oxoglutarate amino transferase), induces the following reaction

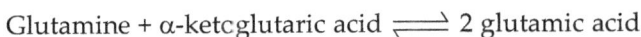

Glutamine + α-ketoglutaric acid $\rightleftharpoons$ 2 glutamic acid

Under these conditions, the glutamine synthase reaction becomes the major route of NH_3 assimilation, i.e., instead of being synthesised by glutamate dehydrogenase,

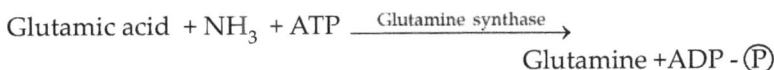

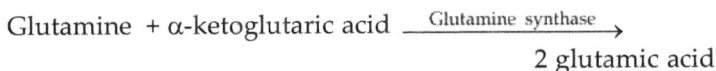

Glutamic acid + NH_3 + ATP $\xrightarrow{\text{Glutamine synthase}}$
    Glutamine + ADP - $\textcircled{P}$

Glutamine + α-ketoglutaric acid $\xrightarrow{\text{Glutamine synthase}}$
    2 glutamic acid

The net reaction is as follows:

α-ketoglutaric acid + NH_3 + ATP → Glutamic acid + ADP + $\textcircled{P}$

In the above reactions, it is clearly noted that the enzyme glutamine synthase–GOGAT utilizes ATP for the reaction activation. But the enzyme glutamate dehydrogenase does not require ATP. In most bacteria, regulatory mechanisms occur, that assure that the glutamine synthase–GOGAT system is utilized when the concentration of ammonia availability to the cell is very low, which reduces the growth rate of the cell were it to be assimilated via glutamate dehydrogenase.

ASSIMILATION OF NITRATE

Nitrate is used by many microorganisms for amino acid and nucleotide biosynthesis and for nitrogen fixation. The assimilation of nitrate is dependent on the state of the compound. Because the valency of the nitrogen atom in NO_3^- is +5, the assimilation of nitrogen from this source involves a preliminary reduction to the oxidation level of ammonia −3.

Some bacteria use nitrate as a terminal electron acceptor for anaerobic respiration. Some algae and fungi do not use nitrate source as such but they extract the nitrogen from nitrate. Hence, they do not undergo the process of anaerobic respiration using nitrate as electron acceptor. But some bacteria do not use nitrogen source. They use the nitrate source. Some organisms, e.g. *Pseudomonas aeruginosa,* use nitrate for both purposes. However, these two processes that reduce nitrate are catalysed by different systems. Assimilatory nitrate reduction is mediated by two enzyme complexes called assimilatory nitrate reductase and assimilatory nitrite reductase

$$NO_3^- \xrightarrow[\text{nitrate reductase}]{\text{Assimilatory}} NO_2^- \xrightarrow[\text{nitrite reductase}]{\text{Assimilatory}} NH_3$$

In the assimilation process where the reduction of nitrate to ammonia by electrons derived from an organic substrate is thermodynamically capable of generating sufficient energy to phosphorylate ADP, in no case is ATP known to be generated as a consequence of nitrate assimilation.

In the assimilation of nitrates in eukaryotes, the nitrate reductase utilize reduced pyridine nucleotides (either NADH or NADPH) as a source of electrons and the electron transfer sequence is thought to be same in all of them.

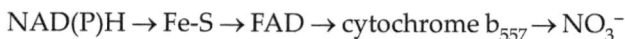

$$NAD(P)H \rightarrow Fe\text{-}S \rightarrow FAD \rightarrow \text{cytochrome } b_{557} \rightarrow NO_3^-$$

In some bacteria, the assimilatory reaction cannot use pyridine nucleotide and the terminal electron acceptor is unidentified. But it is believed that it might be ferredoxin. In all assimilatory reactions, the activation of nitrate reductase reaction is coordinated with molybdenum. Molybdenum acts as a cofactor for this assimilation or dissimilation reaction. But the structure of this molecule is not fully understood yet. It has pterin nucleus structure and is related to folic acid.

Assimilatory reactions in eukaryotes have also been studied completely. However, it seems clear that in all cases the complex 6-electron transfer reaction is catalysed by a single enzyme, and hydroxylamine (NH_2OH) is formed as an intermediate of the process

$$NO_2^- \xrightarrow{\text{Nitrate reductase}} NH_2OH \xrightarrow{\text{Nitrite reductase}} NH_3$$

HOW TO ASSIMILATE MOLECULAR NITROGEN

The availability of nitrogen in the soil and its assimilation are very easy, because the organism in the stagnant environment can easily uptake the nitrogen-containing nutrient by the above mentioned ways. But when present in gaseous state, it requires some special steps for the assimilation process. This can be done through the nitrogen fixation process. The gaseous nitrogen (N_2) with a zero valency must be reduced to ammonia before being inserted into the nitrogenous components of the cell. This fixation process is limited to prokaryotes.

Organisms fix the atmospheric nitrogen either by symbiosis or nonsymbiosis. In the fixation of atmospheric nitrogen, the enzyme nitrogenase is mainly involved. The enzyme has some peculiar properties. (i) It is highly sensitive to oxygen. Hence, at low concentration of oxygen it is inactivated irreversibly (ii) at high concentration of ATP, the enzyme is inactivated. So its ATP requirements must be satisfied by continuous ATP generating system. Nitrogenase enzyme has two enzyme components. They are collectively called as nitrogenase complex Component 1 termed as nitrogenase or MoFe protein and component 2 termed as nitrogenase reductase or Fe protein.

Electrons are transferred through a low potential reductant either ferredoxin or flavodoxin, to nitrogenase reductase. At the end of the transfer reaction with the burst of hydrolysis of molecules of ATP (more than 16 ATP hydrolysed for each molecule of N_2 that is reduced), the electrons are transferred to nitrogenase which helps in the reduction of N_2 and H^+ to NH_3 and H_2. The nitrogenase enzyme has a cofactor in its active site, molybdenum. As indicated previously the structure of the nitrogenase enzyme is a complex one. Various genes are involved in the activation and functioning of the nitrogenase enzyme. It is encoded by the gene called "Nif" gene. Nearly 15 nif genes are arranged in seven continuous operons.

In the nitrogen fixation process hydrogen gas is also produced. This hydrogen gas (H_2) is an inevitable additional product of nitrogen reduction. Loss of this hydrogen gas adds further to the energetic cost of nitrogen fixation. The energy is gained by oxidation of hydrogen with the help of enzyme hydrogenase.

Nitrogenase is highly specific to N_2. But it has some low specificity to some other substrates like N_3, N_2O, HCN, CH_3CN, CH_2CHCN, and C_2H_2. These substrates are also reduced by nitogenase enzyme complex. But reduction of these compounds requires transfer of 2 electrons instead of 6 as in the case of N_2 reduction. The nitrogen fixation in whole cells or in extracts has been greatly provided by the introduction of an assay method using the substrate acelytene, which is reduced to ethylene

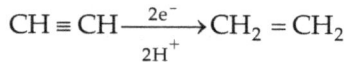

$$CH \equiv CH \xrightarrow[2H^+]{2e^-} CH_2 = CH_2$$

The presence of product can be easily quantitated by gas chromatography, and the reaction is a highly specific one since no system other than nitrogenase can effect this reduction.

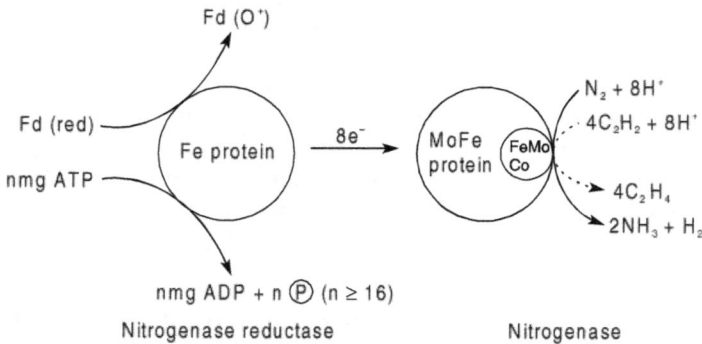

Figure 25.1   Structure and function of the nitrogenase complex

ASSIMILATION OF SULPHUR

Many sulphur-utilizing microorganisms satisfy their sulphur requirements from sulphur. Sulphur in the oxidized state, has 6 valency which is reduced to sulphide (valency-2) prior to its incorporation into cellular organic compounds. Sulphur is reduced by sulphate-reducing bacteria during anaerobic respiration. The terminal acceptor for anaerobic respiration is sulphur. The activation of sulphur assimilation and the enzymes involves in this pathway are quite different. The reduction of sulphate from sulphate source for use is termed as assimilatory sulphate reduction (analogous with assimilatory nitrate reduction).

PATHWAY OF ASSIMILATORY SULPHUR REDUCTION

In the assimilatory reduction reaction, activation of sulphur takes place by the reduction of two electrons and thus the sulphur has converted to its activated form, adenyl sulphate with the help of the enzyme sulphate adenyl transferase. Further conversion requires three enzymes with the expenditure of three high-energy phosphate bonds. The final six electron reduction is catalysed by a huge, complex flavometalloprotein, sulphate reductase. The molecular weight of the sulphate reductase in *E. coli* is 750,000 and contains 4 FAD, 4 FMN, and 12 Fe prosthetic groups.

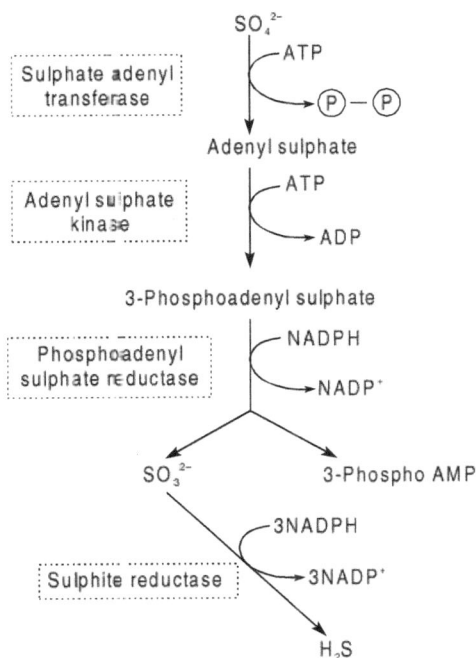

Figure 25.2 Assimilatory reduction of sulphate to produce H_2S for use in biosynthetic reactions.

STUDY OUTLINE

- Nitrogen and sulphur can be utilized by microorganisms in assimilatory reactions.

- Nitrogen is the major source for amino acid and nucleotide biosynthesis. It is converted as glutamine or glutamic acid in microorganisms. Glutamine is the mother molecule for amino acid biosynthesis.

- Microorganisms convert this molecular nitrogen and fix the nitrogen for their metabolism.

- Nitrogen is fixed by a special type of enzyme called nitrogenase enzyme which is highly sensitive to oxygen.

- Nitrogenase has two enzyme components including nitrogenase and nitrogenase reductase.

- Molybdenum acts as a cofactor for the enzyme nitrogenase activation.

- In the case of sulphur, it is also assimilated into hydrogen sulphide, and sulphur act as a major source for sulphide bond formation in amino acid biosynthesis.

CONCEPT CHECK

1. Give a detailed account on assimilation of nitrate?

2. How do organisms fix the atmospheric nitrogen?

3. Give an outline about the assimilation of sulphur?

CRITICAL THINKING

As far as we know from this chapter, the nitrogenase enzyme is highly sensitive to oxygen. But the organism takes up oxygen for various purposes, and certain metabolic pathways also need oxygen. In the presence of oxygen in the system, how is the nitrogenase enzyme protected by the microorganisms?

26

NITROGEN FIXATION

Nitrogen gas occupies 80% of the earth's atmosphere. The reduction of the molecular atmospheric nitrogen into ammonia takes place in only few systems. Some free-living microorganisms like *Azotobacter, Klebsiella* and *Cyanobacteria* fix the atmospheric nitrogen independently. They are called as "non symbiotic" microorganisms and the process is called as "nonsymbiosis". But in case of *Rhizobium*, it fixes nitrogen by forming symbiotic nodules on the roots of the leguminous plants like bean or alfalfa. These are called "symbiotic" microorganisms and the process is called as symbiosis.

Symbiotic microorganisms fix nitrogen by forming a specialized structure in their nodules "Bacteroid ." Some trees, such as alder, also form nitrogen fixing nodules and thus have the capacity to fix nitrogen. The nitrogen is fixed by the help of the enzyme nitrogenase which has high affinity for triple bonds.

For example the laboratory experiments revealed that nitrogenase has high affinity to acetylene with which it forms ethylene as follows.

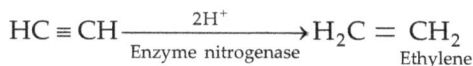

$$HC \equiv CH \xrightarrow[\text{Enzyme nitrogenase}]{2H^+} H_2C = CH_2 \quad \text{Ethylene}$$

The comparable reaction with nitrogen is

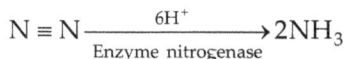

$$N \equiv N \xrightarrow[\text{Enzyme nitrogenase}]{6H^+} 2NH_3$$

This conversion of molecular nitrogen to ammonia is used in the manufacture of ammonia-based fertilizers. Interest in the molecular details of biological nitrogen fixation has derived partly from hopes of supplanting their energy-intensive process with a means of ammonia production that can take place under milder conditions.

When we compare photosynthesis and N_2 fixation, it shows that both N_2 and CO_2 are stable inorganic compounds whose reduction requires both energy and low potential electrons, i.e. electron carriers of very low E_o. However, photosynthesis uses light to generate both energy (through photophosphorylation) and low potential electrons (as ferredoxin). The major reason for fixing nitrogen by *Rhizobium* in leguminous plants is the abundant presence of the protein called leg hemoglobin, which maintains an anaerobic environment by binding any O_2 that finds its way into the nodule and presenting it to respiratory enzymes.

NITROGENASE ENZYME

The nitrogenase enzyme complex has been characterised as two components and neither is achieved without the other. Component I is nitrogenase and component II is nitrogenase reductase.

Component I is otherwise called as MoFe protein (Mo for molybdenum, Fe for iron). Component II, which is a smaller molecule is designated as the Fe protein. Both molecules contain sulphur. Component I catalyses the reduction of N_2, and component II transfers electrons from ferredoxin or flavodoxin to component I. Both proteins contain iron-sulphur clusters, and component I also contains molybdenum, in the form of a tightly bound iron–molybdenum cofactor (FeMo-Co). N_2 binds to this cofactor during its reduction although the precise mode of binding is not yet known.

Nitrogenase enzyme complex consist of

i.   Component I and Component II

ii. A strong reducing agent such as ferredoxin or flavodoxin

iii. ATP

Crystal structure reveals that the entire complex was reported with a nonhydrolysable ATP analog bound to component II. So the binding of ATP evidently forces a conformational change in component II which drives its docking to component I. Component I contains 2 novel iron–sulphur complexes—the P cluster, which contains 7 sulphur and 8 irons, and the iron–molybdenum cofactor which contains 9 sulphur, 7 irons, and one molybdenum ion linked also to a molecule of homocitrate.

$$CH_2-COO^-$$
$$|$$
$$CH_2$$
$$|$$
$$HO-C-COO^-$$
$$|$$
$$CH_2-COO^-$$

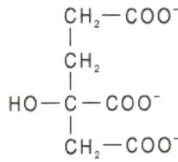

Structure of homocitrate

Electrons flow from ferredoxin or flavodoxin to the Fe_4S_4 complex in component II, and the hydrolysis of bound ATP somehow drives

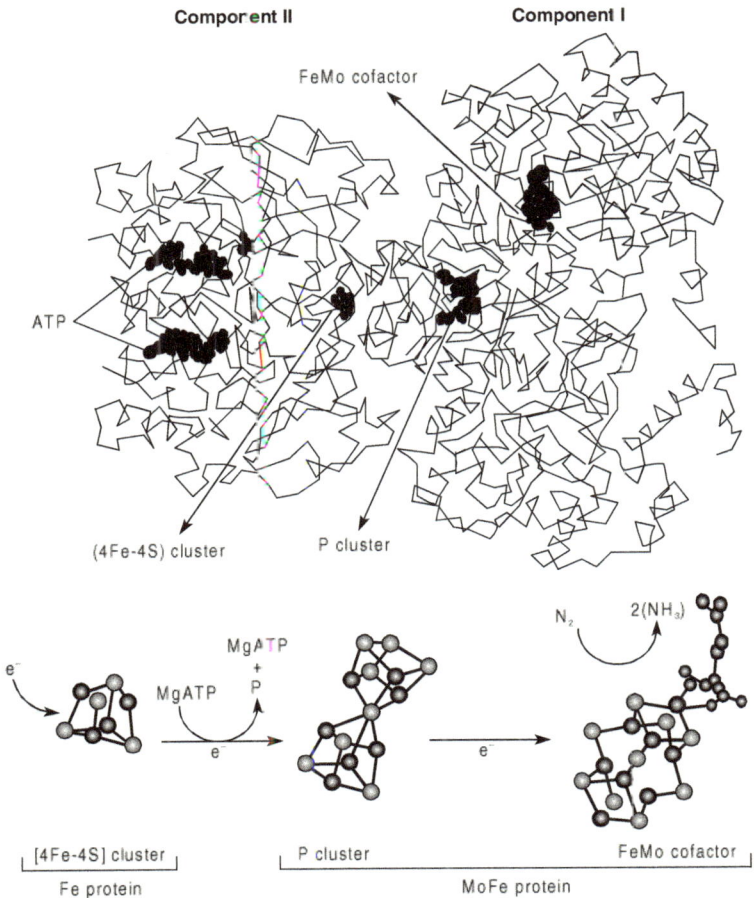

Figure 26.1   Structure of nitrogenase enzyme.

the electrons to the P cluster in component I and then to FeMo–Co. These three clusters are sufficiently close together in the complex to allow facile electron transfer. Almost certainly, the reduction of N_2 occurs at the bound coenzyme, but the mode of N_2 binding must be described in order for the mechanism of reduction to be understood.

iv.   A regulating system for NH_3 production and utilization

v.   A system that protects the nitrogen fixing system from inhibition by molecular oxygen.

Mechanic details of nitrogen fixation appear quite similar among the species examined to date. As studied in *Klebsiella pnenmoniae*, the stoichiometry of the overall reaction is as follows.

$$N_2 + 8e^- + 16ATP + 16H_2O \rightarrow 2NH_3 + H_2 + 16ADP + 16 \, Pi + 8H^+$$

NONSYMBIOTIC NITROGEN FIXATION

Nonsymbiotic nitrogen fixation has been studied in two bacterial species—*Clostridium pasteurianum* and species of *Azotobacter*. These bacteria are the only ones which fix the atmospheric nitrogen in free-living condition. The former are the anaerobic rod-shaped bacteria and the latter are aerobic oval or spherical cells; both are widely distributed in soils. The capacity of nitrogen fixation by *Azotobacter* species is greater than that of *Clostridrum pasteurianum*. It has been estimated that the amount of nitrogen fixed by the nonsymbiotic process ranged between 20—50 lb/acre annually.

SYMBIOTIC NITROGEN FIXATION

This type of nitrogen fixation is found in *Rhizobium* species. These organisms fix nitrogen by forming nodules in the plant root. They harbour mainly the dicot plants. Before fixation they have to establish in the root. So they form infection threads in the root hairs and invade the root through these infection threads. After invasion they cause the enlargement of the plant and rapid cell division of the plant cell leads to the abnormal growth (nodules) on the root. In this association both the living systems are benefited. The organism fixes the atmospheric nitrogen in the root which can be used by the plant, in turn the plant roots provide nutrients for the growth of the microorganisms.

Only some species of *Rhizobium* produce nodulation and nitrogen fixation in any legume. There is a degree of specificity between the bacteria and legumes. *Rhizobum* species or strains effective for one group of plants are less effective or ineffective for other groups. Even within a species, certain strains are more effective than others within a given host plant.

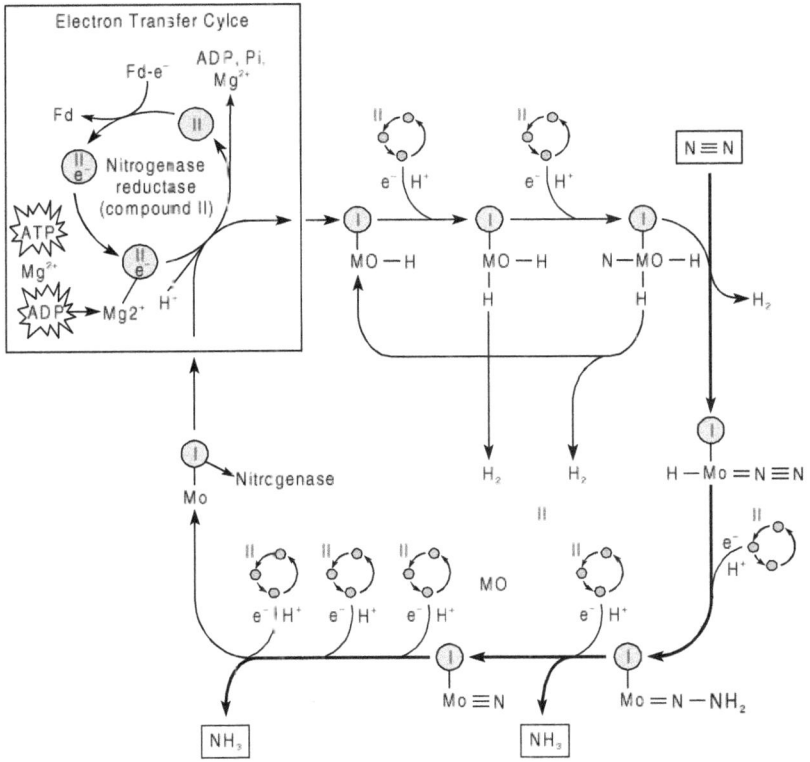

Figure 26.2   Diagrammatic representation of nitrogen fixation.

STUDY OUTLINE

• Nitrogen fixation is nothing but the fixation and conversion of atmospheric molecular nitrogen to ammonia.

• Nitrogen can be fixed either by symbiotic or nonsymbiotic ways.

- The nonsymbiotic fixers like *Azotobacter* and *Klebsiella* are present freely in the soil.

- But symbiotic fixers are found in associated form, e.g. *Rhizobium*. They form nodules in the roots of legume plants and thus fix the nitrogen.

- The enzyme nitrogenase is responsible for the fixation of atmospheric nitrogen.

- Nitrogenase has two component systems—nitrogenase or component I and nitrogen reductase.

- Nitrogenase enzyme has high affinity on triple bonds.

- Nitrogenase is very sensitive to oxygen. The protein leg hemoglobin is responsible for the anoxygenic condition in nodules.

- The bacteria which forms a specialized structure in the nodule for the fixation of nitrogen is called bacteriod.

CONCEPT CHECK

1. Give a detailed account of symbiotic and nonsymbiotic nitrogen fixation?

2. How does the nitrogenase enzyme fix the atmospheric nitrogen?

3. Write about the nitrogenase enzyme complex.

CRITICAL THINKING

1. Why does Rhizobium particularly fix the nitrogen in legume plant?

2. Rhizobium is a soil gram-negative bacterium. How is this bacterium attracted towards the legume plant root? Is any chemical substance responsible for this attraction?

27

FERMENTATION

INTRODUCTION

Fermentation is an energy-yielding metabolism that involves a sequence of oxidation—reduction reactions. In other words we can say that it is the conversion or breakdown of complex molecules into simpler ones by the enzymes in a sequence of enzymatic reactions. In fermentation, both the acceptor and donor are internal i.e., they are derived from the same organic substrate which is involved in the fermentation process. In fermentation pathway, the oxidation state of the substrate is not changed. The oxidation products are counterbalanced by reduced products and thus the oxidation and reduction reactions are evenly maintained. Coenzymes involved in this are reduced and after that they are reoxidised and thus not consumed by the reaction.

In fermentation, ten ATP are released. But in respiration, more energy is released. This is because the organic substrates involved in fermentation can act both as acceptor and donor. Thus the oxidation of carbon and hydrogen into CO_2 and H_2O will not take place. But the molecules are just rearranged. However, in respiration, the oxidation of carbon and hydrogen readily yields CO_2 and H_2O which splits and releases more energy which is trapped within ATP. This can be explained with one example.

When glucose is oxidized during metabolism the ΔG° value for the oxidation of glucose to CO_2 is −686 Kcal/mol. But when glucose is partially fermented (oxidized) to two molecules of the fermentation product, lactic acid, the ΔG° value is only −58 Kcal/mol.

This clearly indicates that during respiration (oxidation) more energy is trapped than during fermentation.

In fermentation more substrates are utilized than in respiration. Hence respiration is more beneficial than fermentation with regard to both energetics and utilization of available organic nutrient source. But organisms undergo fermentation only under unfavourable conditions of metabolism, i.e., if they lack the external acceptor or donor. Some organisms which are sensitive to oxygen can synthesize their ATP through the fermentation process.

Fermentation in other terms, is named as anaerobic metabolism, because it takes place only in the absence of oxygen. Even though the organism is aerobic, it cannot use the molecular oxygen. This is mainly carried out in anaerobic organisms. In the fermentation pathway lesser ATP is synthesised by substrate level phosphorylation. But the respiration pathways are restricted to this type of phosphorylation. In case of respiration, high amount of energy is synthesised through the oxidative pathway and chemiosmosis.

Generally ATP is synthesized through oxidative phosphorylation with the help of the enzyme FoF_1, ATPase. But this is not found in fermentative organisms. Obligate anaerobic bacteria like *Streptococcus* have this FoF_1 ATPase in their cytoplasmic membrane. In such bacteria, ATP is used to pump protons through the FoF_1 complex in the reverse direction. In this process, ATP is synthesized by substrate-level phosphorylation in fermentation and the Pi + ADP is converted into ATP with the help of the enzyme ATPase. This reaction is also coupled with the export of protons from the cell. The FoF_1-ATPase system thereby generates a proton motive force across the cytoplasmic membrane. This helps in the maintenance of intracellular pH at the appropriate value and provides a mechanism for driving processes that depend on the proton motive force across the membrane, such as the active uptake of sugars and other substrates, the export of Na^+ and Ca^{2+}, and the rotation of flagella.

LACTIC ACID FERMENTATION

Various substrates undergo various types of fermentation. They are homolactic acid, heterolactic acid, ethanol, mixed acid, propionic acid, butanediol, butyric acid, amino acid and methanogenesis. Among these homolactic and heterolactic acids are found in various microorganisms. The organisms which carry out this type of fermentation are called as lactic acid bacteria e.g., *Streptococci* and

Lactobacillus. The organism converts the pyruvate substrate into lactic acid with the coupling reaction of reoxidation of NADH and formation of NAD⁺.

Homolactic Fermentation

In homolactic fermentation the only end product formed is lactic acid. The Embden–Meyerhof pathway of glycolysis is used in the lactic acid fermentation pathway. The organisms involved in homolactic fermentation are *Streptococcus, Lactococcus, Lactobacillus, Enterococcus, Pediococcus,* etc. This type of fermentation is necessary for industrial production purposes, particularly in dairy industry for the preparation of sour milk, and the production of cheese, yoghurt, and various other dairy products.

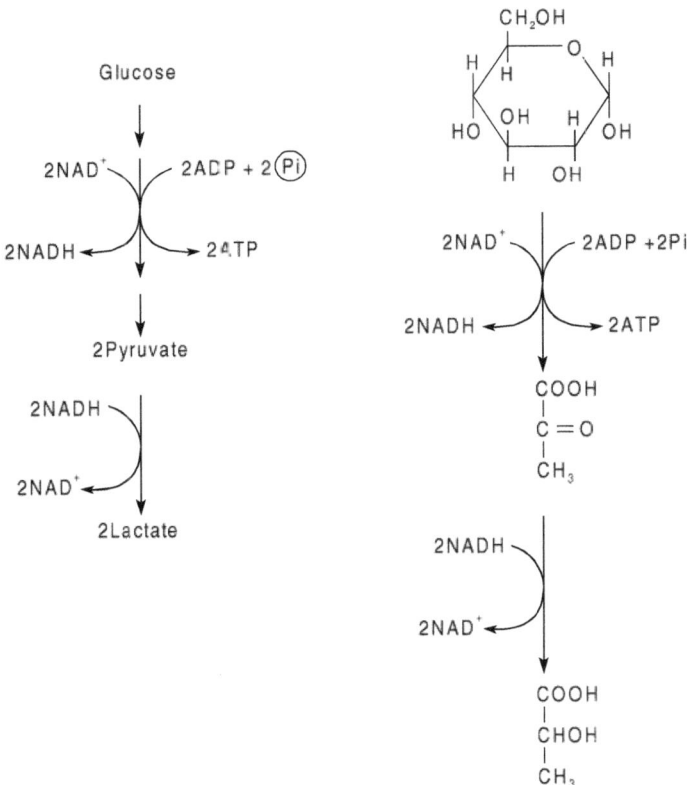

Figure 27.1 Homolactic acid fermentation pathway. Production of lactate from pyruvate

Streptococcus species harbour on tooth surfaces or oral cavity and produce lactic acid by homolactic fermentation which causes dental plaque and gradually eats through the enamel of the tooth, creating caries (cavities). *Lactobacillus* is present in the human digestive tract and mediates the digestion of milk. The organism colonizes the intestinal tract and so are called as initial colonizers. *Lactobacillus acidophilus* is added to commercial milk products, and given to patients who are not able to digest milk due to lack of this organism. The organism produces enzymes which convert the milk and thus digests them, so they prevent the accumulation of undigested milk which causes gastrointestinal problems.

Heterolactic Fermentation

Some microorganisms produce lactic acid through heterolactic acid pathway, because in addition to lactic acid, ethanol and carbon dioxide are also produced. The organism proceeds to this type of fermentation by pentose phosphate pathway but homolactic fermentation, proceeds through the Embden–Meyerhof pathway. The glycolytic portion of this pathway yields ethanol and CO_2.

Heterolactic fermentation can be expressed in the following equation

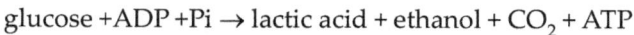

$$\text{glucose} + ADP + Pi \rightarrow \text{lactic acid} + \text{ethanol} + CO_2 + ATP$$

This type of pathway is carried out in *Leuconostoc* species. In this pathway, only one ATP is produced for each molecule of glucose. The organism is used in sauerkraut production. Some other *Lactobacillus* species are also used for this purpose.

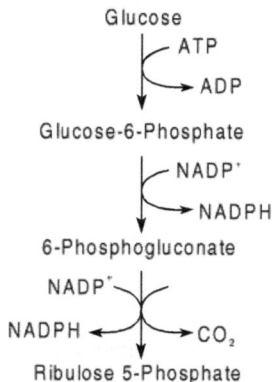

Glucose

$\llcorner$ ATP

$\llcorner$→ ADP

Glucose-6-Phosphate

$\llcorner$ NADP·

$\llcorner$→ NADPH

6-Phosphogluconate

NADP·

NADPH ← → CO_2

Ribulose 5-Phosphate

Contd.

Xylulose 5-Phosphate

CoA

Acetyl CoA   Glyceraldehyde 3-Phosphate

NADH   NAD⁺
NAD⁺   NADH
         2ADP + 2 (Pi)
Acetaldehyde   2ATP

NADH   Pyruvate
NAD⁺   NADH
Ethanol   NAD⁺

Lactate

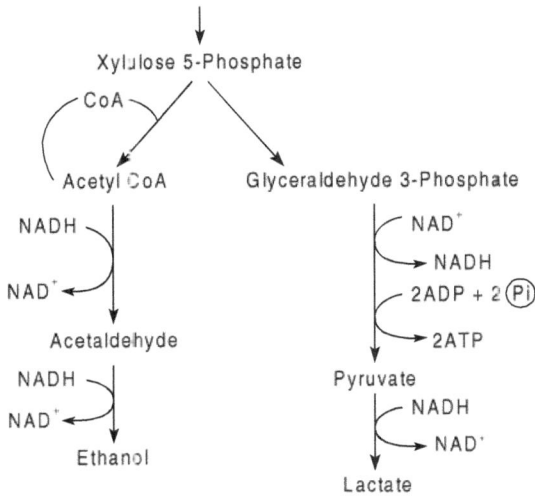

Figure 27.2 Heterolactic acid fermentation by *Leuconostoc* species.

ETHANOLIC FERMENTATION

The name ethanolic fermentation is so given because ethanol is the major product in this fermentation pathway. By the conversion of glucose, the pathway yields ethanol and CO_2. The terminal reaction in ethanolic fermentation is coupled with the conversion of NADH to NAD^+. If glucose is the substrate for ethanolic fermentation, the overall reaction in Embden Meyerhof pathway is

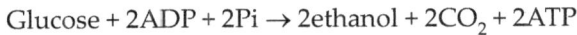

$$\text{Glucose} + 2ADP + 2Pi \rightarrow 2\text{ethanol} + 2CO_2 + 2ATP$$

Ethanolic fermentation is economically important as it is mainly used in food and industrial microbiology. The organism which carries this type of fermentation is *Saccharomyces cerivisiae*, which includes baker's and brewer's yeast.

They are used in the production of alcoholic beverages like wine, beer, and also for the production of distillery spirits. Among these, the most useful production is gasohol, which is used as a fuel. Ethanolic fermentation is also used in bread making, because during fermentation carbon dioxide is released which makes the bread soft by making it rise.

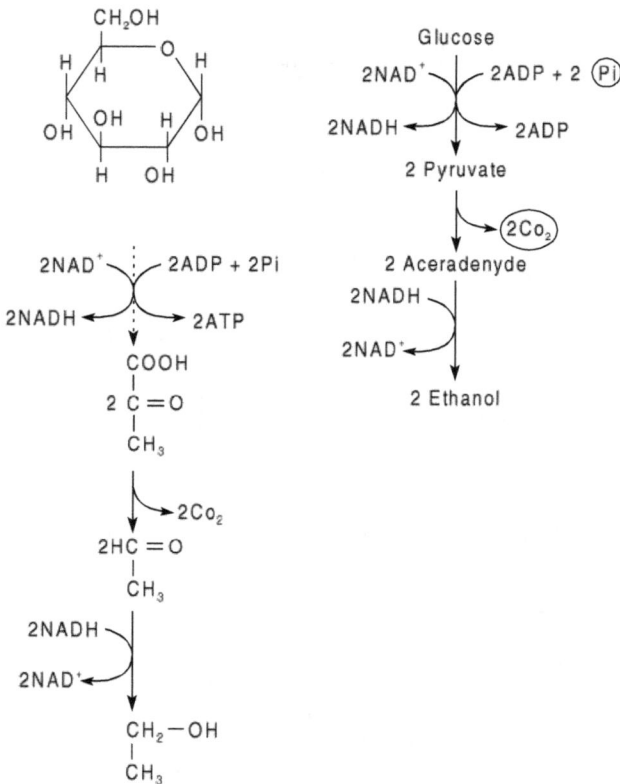

Figure 27.3   Ethanolic fermentation by *Saccharomyces cerevisiae* through Embden–Meyerhof pathway. CO_2 obtained as additional product.

PROPIONIC ACID FERMENTATION

This type of metabolism is carried out in propionibacterium which produces propionic acid from carbohydrates and lactate. It is a gram-positive, rod-shaped organism. Some propionibacteria utilize lactate as their substrate for fermentation. This organism is used in late stages of fermentation. i.e., during cheese production, *Lactobacillus* is used for the conversion of substrate to lactic acid and after that propionibacteria is used for the conversion of lactic acid to propionic acid and carbon dioxide. Propionibacteria produce CO_2 which can be identified by the bubbles in cheese or curd. In swiss cheese, characteristic bubbles which form holes on the surface are produced by proponibacteria and this gives additional flavour to the swiss cheese.

Mixed acid Fermentation

This type is so named because the end product is a mixture of substances. This type of fermentation is carried out in species of the Enterobacteriaceae family like *E. coli, Shigella* and *Salmonella*. These organisms produce various products including ethanol, acetate, succinate, formate, molecular hydrogen, lactate and carbon dioxide by the conversion of pyruvate. But the overall proportion of the CO_2

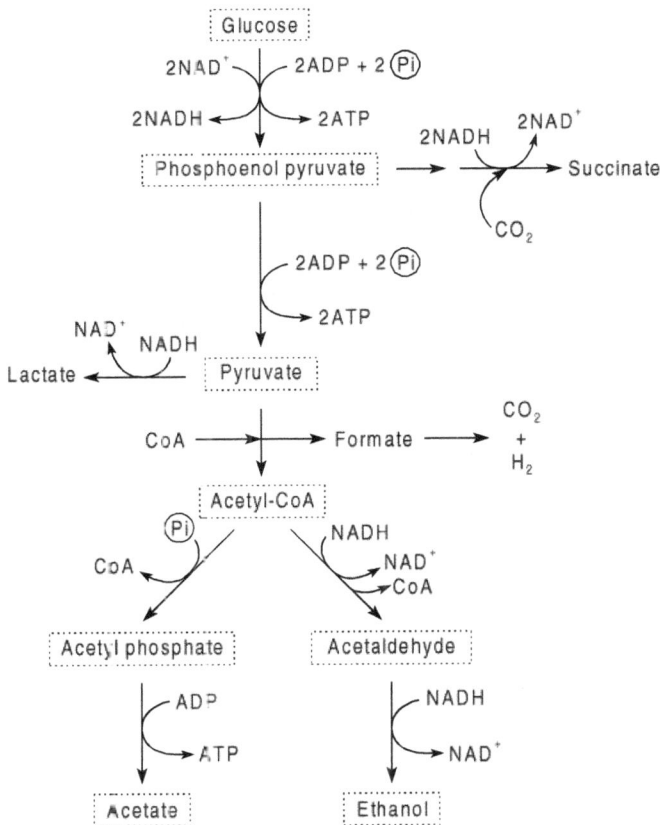

Figure 27.4   Mixed acid fermentation pathway and the production of CO_2 by *Escherichia coli.*

and H_2 is equimolar because of the presence of the enzyme formic hydrogen lyase of the pyruvate–formate lyase system. The reaction produces equal amount of CO_2 and H_2 as follows:

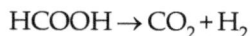

$$HCOOH \rightarrow CO_2 + H_2$$

In addition to that, the reduction of NADH and reoxidation of NAD$^+$ takes place. The formation of acetate also produces ATP through substrate level phosphorylation. The presence of acid end products can be analysed by biochemical tests. The test for the analysis of acid end product is methyl red (MR) test. Methyl red is an acid–base indicator which gives cherry red colour in acidic condition. The presence of cherry red colour indicates the presence of mixed acid end product of the organism. This is because the concentration of acidic end products is five times larger than the neutral end products. This method can be used to identify the microorganism in the laboratories and also for clinical diagnosis.

BUTANEDIOL FERMENTATION

In the production of neutral end product by fermentation pathway, the neutral end product is butanediol. So the name given is Butanediol pathway because some organisms produce larger amount of neutral end products. This fermentation pathway is carried out by *Enterobacter*, *Serratia*, *Erwinia* and *Klebsiella* species. The presence of neutral end product can be checked by the intermediate substances like acetoin (acetyl methyl carbinol). The test is performed to find the presence of acetoine is Voges-Proskauer (VP) test. This test is generally used with MR test to differentiate the organisms like those of the family Enterobacteriaceae, based on their biochemical characters. Among other reasons, the ability to make this distinction is very important because *E. coli* is used as an indicator of human faecal contamination in processes that assess the safety of water supplies.

Some organisms like *Klebsiella* produce both acid and neutral end products. But they show a positive result to VP and a negative one to MR. This is because they can produce enough amount of acetoin to perform the test but not the acidic end products.

The typical ratio of neutral (butanediol and ethanol) products to acidic products (acetate, formate, lactate and succinate) in an organism carrying out both the butanediol and mixed acid fermentation

CH₂OH

Glucose

NAD⁺ ⟍ ADP + Ⓟⁱ
NADH ⟍ ATP

2 Pyruvate

NAD⁺ ⟍ ADP + Pi
NADH ⟍ ATP

2NADH
2NAD⁺
CO₂ + H₂ + Ethanol
CO₂

CH₃
2C = O
COOH

Acetolactale

CO₂

H₂ + CO₂ ← COOH
CH₃CH₂OH
2NAD⁺
CO₂

Acetoin

NADH
NAD⁺

CH₃
C = O
HO — C — COOH
CH₃

2, 3 Butanediol

CO₂

CH₃
C = O
H — C — OH
CH₃

NADH
NAD⁺

CH₃
CHCH
CHCH
CH₃

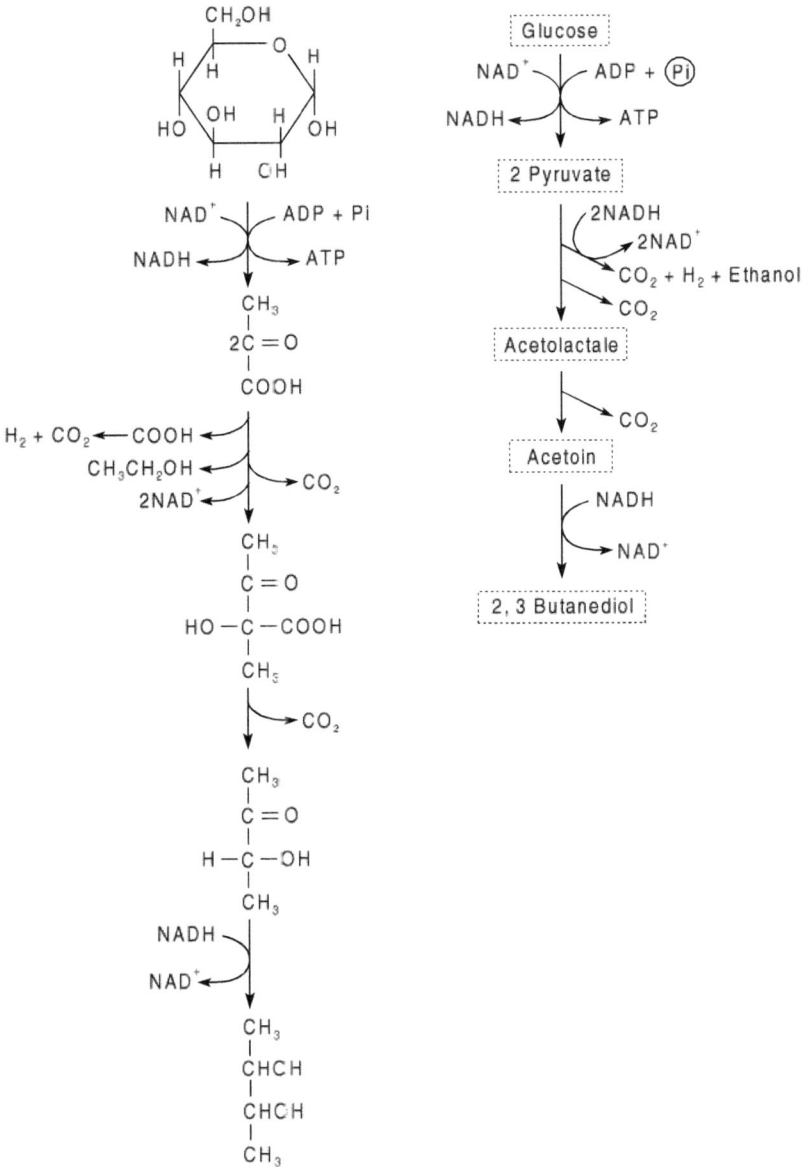

Figure 27.5 Production of butanediol by *Enterobacter* spp. from butanediol fermentation.

pathway are 6 : 1. CO_2 and H_2 are mixed and also produced by such organisms, typically in the ratio of about 5 : 1, since CO_2 is produced at several steps and H_2 is produced only from the decomposition of formate.

Butyric Acid Fermentation

Butanol fermentation pathway was first discovered by Chaim Weizmann, and was particularly important because it allowed Britain to produce acetone for use in the manufacture of ammunitions during World War I.

Butyric acid fermentation is otherwise called as butanol fermentation which is carried out by *Clostridium* species. The conversion of pyruvate yields either acetone and carbon dioxide, isopropanol and carbon dioxide, butyrate or butanol. The production of butanol is very useful in commercial aspects. The acetone from this fermentation is used as nail polish remover. So today, the choice of using microbial or organic–synthetic means to produce solvents is based on economic factors.

AMINO ACID FERMENTATION

In addition to carbohydrate metabolism, various pathways are involved in extracting the carbon and energy source of microorganisms. Thus some microorganisms use amino acids to satisfy their carbon and energy requirements. For example some anaerobic microorganisms like *Clostridium*, ferment arginine. With the help of arginine dihydrolase enzyme, arginine is converted to ornithine, CO_2 and NH_3 (ammonia). In the first step arginine is deaminated to citrulline, which is then phosphorylated and split into ornithine and carbamyl phosphate. Carbamyl phosphate has high energy phosphate bond which can be utilized in substrate level phophorylation for the synthesis of ATP.

Another organism, *Peptococcus anaerobius* leads to the formation of acetyl phosphate, which is then utilized in a substrate level phosphorylation to generate ATP. This reaction takes place in the fermentation of glycine to acetate. Some other amino acids that can be fermented by various anaerobes, although without the additional synthesis of ATP, are threonine, glutamine, lysine, and aspartate.

Some microorganisms carry out mixed amino acid fermentation, e.g., *Clostridium*. During extensive protein degradation, the reaction involves one amino acid as electron donor and one amino acid as electron acceptor in an oxidation–reduction reaction. This type of coupling oxidation–reduction between the two amino acids is called the "Stickland Reaction". The deamination and decarboxylation of the amino acids takes place mainly in this fermentation. For example a mixture of alanine and glycine can yield the end products of acetate, carbon dioxide and ammonia. The mixed amino acid fermentation pathway gives the pleasant smell of some wines and cheese but is also partly responsible for the unpleasant odour of "Gangrenous" wound.

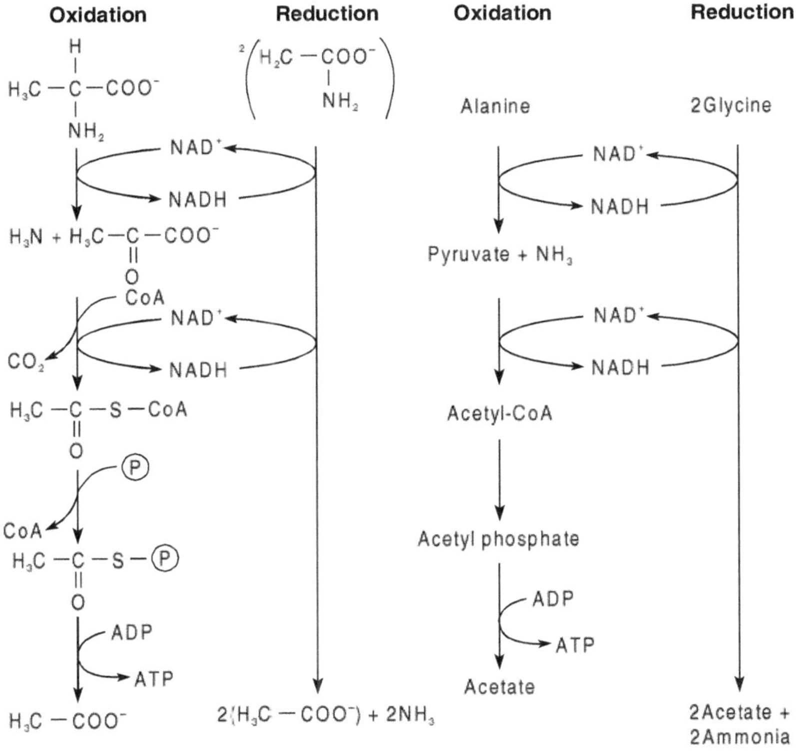

Figure 27.6 Oxidation–reduction reaction (Stickland Reaction) of two amino acids, alanine and glycine, mediated by *Clostridium sporogenes*.

METHANOGENESIS

Some anaerobic microorganisms produce methane in a type of metabolism called as methanogenesis and the organisms involved in this type of metabolism are called methanogens. Mostly these organisms are chemoautotrophs, they satisfy their electron requirements from molecular hydrogen and formate which forms CO_2 and CH_4 (methane). Some microorganisms produce methane from acetate and they are called chemoorganotrophs. On earth, 1/3 of the methane is produced by chemoautotrophs and 2/3 of the methane is produced by chemoorganotrophs.

In the acetate fermentation of methanogenesis, the methyl group is reduced to methane and the carboxyl group is oxidized to carbon dioxide. The reaction is as follows.

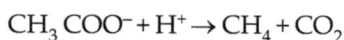

$$CH_3 COO^- + H^+ \rightarrow CH_4 + CO_2$$

This methane formation pathway from acetate begins with the reaction of acetate and coenzyme A, which forms acetyl-CoA. This reaction burns with the help of the fuel ATP. Acetyl-CoA then cleaves into CoA with the help of the nickel/cobalt-containing enzymes and the formed acetyl-CoA is oxidized to CO_2 and the remaining methyl group is attached to 2-mercapto ethane sulphonic acid (HS–CoM). The resulting CH_3–S–CoM is subsequently reduced with two electrons from 7-mercaptoheptanoylthreonine phosphate (HS–HTP) to form methane and a disulphide of CoM–S–S–HTP. ATP is generally utilized in this reaction. However, the archaea produce their ATP by the formation of high chemical potential across the cytoplasmic membranes of methanogens, and it is quite likely that electron transport and chemiosmosis generation of ATP occurs in these archaea.

STUDY OUTLINE

- Fermentation is simply defined as the conversion of complex substances into simpler substances with the help of enzymes, with the release of energy.

- Various types of fermentation include lactic, acetic, butyric, and mixed acid fermentation.

- The surprising point in fermentation is that both the electron donor and acceptor is the same organic compound.

- Fermentative products are mainly used in production of wine, bread, cheese, and some other commercial production.

- Depending on the end product, lactic acid fermentation is classified into homo-and heterolactic acid fermentation.

- Some microorganisms convert acetate to methane through the fermentative pathway.

CONCEPT CHECK

1. Explain the fermentative pathway of *Streptococcus* and *Lactobacillus*.

2. Explain the mixed acid fermentative pathway.

3. What is the use of adding methyl red and Voges Proskaur in biochemical analysis?

4. Give a detailed account of homo-and heterolactic fermentation.

5. How does *Clostridium* ferment amino acids?

6. What is Stickland reaction?

7. Write about the commercial uses of various fermentation processes.

CRITICAL THINKING

1. Most of the microorganisms ferment through Embden–Meyerhof pathway. Is there any probability of microorganisms to carry out fermentation through the Entner–Doudoroff pathway?

2. Most of the methanogens are strict anaerobes. How do they oxidize their substrate for the production of carbon dioxide?

28

REPRODUCTIVE PHYSIOLOGY OF FUNGI AND BACTERIA

REPRODUCTIVE PHYSIOLOGY OF FUNGI

The study of fungi is called mycology. Fungi include yeasts, molds and fleshy fungi. Yeasts are unicellular organisms. Molds are multicellular filamentous organisms, e.g. mildews, rusts and smuts. Fleshy fungi are multicellular, filamentous organisms that produce a thick (fleshy) reproductive body. In fungi, three kinds of reproduction have been observed so far.

1. Vegetative reproduction
2. Asexual reproduction
3. Sexual reproduction

Vegetative Reproduction

In this reproduction, a piece of mycelium (collection of hyphae), when broken off, is capable of giving rise to a new mycelium. This method of reproduction is called as "Fragmentation".

Asexual Reproduction

This type of reproduction takes places by means of unicellular reproductive bodies called spores. When the mycelium is few days old, some hyphae grow upward and the tip of each swells up to form a spherical sac called "sporangium". The vertical hyphae which bear the sporangia are called "sporangiophores". They occur singly or in clusters depending on the organism. The contents of the sporangiophore become dense at the tip and then a cross septum is formed. The tip later becomes dome shaped and projects into sporangium to form "columella". The cytoplasm within the sporangium breaks up into a number of small pieces, each with a nucleus to form "spores." When the spores are mature, the sporangium bursts and sets the sproes free. Many of the spores get destroyed since they do not come in contact with favourable conditions. But those that happen to settle on a suitable medium like damp dung, bread, jam, pickles or other organic matter, germinate and form new mycelia.

Sexual Reproduction

This type of reproduction takes place only between two hyphae of different mycelia. When the tip of two such hyphae come in contact with each other, a small bit of the tip of each hypha gets separated by means of a septum to form a gametangium. These gametangia fuse together to form a zygospore which surrounds itself with a thick, black, warty wall consisting of two layers—an outer thick highly cuticularized covering called exosporium and an inner thin delicate "endosporium". Later when the zygospore germinates, the exosporium ruptures and the endosporium grows out in the form of a hyphae called promycelium, which bears a single spherical sporangium at its tip. The sporangium produces a number of spores, each one of which germinates into a mycelium.

REPRODUCTIVE PHYSIOLOGY OF BACTERIA

Bacteria are very small, simple and single-celled (unicellular) organisms whose genetic material is not enclosed in a special nuclear membrane. Among bacteria, reproduction is usually an asexual process. After the cell has reached its maximum size, it splits at right angles to the long axis to form two new cells. This is called as "binary fission" or "simple transverse fission."

Binary Fission

In this type of reproduction, the protoplasm of the bacterial cell is first divided into two, more or less equal parts by the growth of the transverse septum from the plasma membrane and cell wall. In some species the septum thus formed splits into two and the daughter cells separate almost immediately. In other forms, the cell wall of the daughter cell remain continuous and the organisms form characteristic cell groups. Under favourable conditions, growth and reproduction are repeated with great rapidity, i.e. every twenty to thirty minutes so that one individual cell may reproduce millions of new organisms in less than a day. Until recently, binary fission was the only kind of reproduction known among bacteria. Now it is known that in some bacterial species, there can be conjugation of two individuals with genetic recombination. It was first identified by Lederberg and Tatum in 1946.

Conjugation

A clear understanding of conjugation in bacteria came about with the discovery that there is sexual differentiation in *E. coli*. In other words, different mating types of the bacteria exist. Conjugation occurs between two closely related but different bacterial cells, a donor and recipient. The donor cell known as the male, contains a conjugation plasmid with an F factor. The recipient or female cell does not contain these factors. As a result of this difference, the male is often called as F+ cell, and the female cell is termed F- cell. The 'F' factor thus determines the maleness of the cell. Like other plasmids the F factor can be lost from the cell while growing. F cells incubated in the stationary phase of growth for prolonged periods are cured of their F factor and become changed to F- cells. F+ cells contain not only the sex factor but also have a special elongated pilus known as the "sex pilus" the production of which is determined by genes of the 'F' plasmid. If F+ cells are mixed with F cells in a culture the two cells come in contact with each other through the F pilus. During that time the one strand of the F plasmid is nicked at a site called oriT (origin of transfer), and the 5' end of the strand is transferred into the recipient. During transfer the plasmid is replicated via a rolling circle mechanism of replication. Inside the recipient cell the donor copy is recircularized (a completed copy of the donor remains intact). Now the recipient cell contains plasmid like donor and becomes F+. Genetic markers other than sex factor located on the F plasmid are also transferred to the recipient cell.

STUDY OUTLINE

- There are three kinds of reproduction that have been observed so far—(i) vegetative reproduction (ii) asexual reproduction (iii) sexual reproduction.
- Vegetative reproduction is otherwise called as fragmentation.
- Bacteria can be reproduced by binary fission or simple transverse fission.

CONCEPT CHECK

1. Explain the reproductive physiology of fungi.
2. Give an outline concept of bacterial reproduction.

APPENDIX

LABORATORY EXPERIMENTS

Important Safety Guidelines

1. Wear clean laboratory coats while working in the laboratory.

2. Do not eat or drink in the laboratory.

3. Do not use laboratory glassware for drinking or eating purposes.

4. Keep your fingers, paper, pencil, etc. out of your mouth.

5. Keep clean as possible while handling the equipments such as microscopes, centrifuges, etc.

6. Use good permanent glass marker for labelling on the glassware to be autoclaved or to be kept in water bath.

7. In case any culture is spilled, inform your instructor immediately and follow the instructions for cleaning. After cleaning, wash your hands with soap.

8. At the end of your experiment replace the equipment at their respective places.

9. Clean the table tops with disinfectant before and after work.

10. At the end of the work, all cultures should be discarded after inactivation according to the instructions given.

11. Agar and liquid waste should be displaced only in the closed box provided for the purpose.

12. Wash your hands with soap before and after work in the laboratory.

Experiment 1 Preparation of Liquid Media for Cultivation of Microorganisms

For the cultivation of microorganisms liquid media (Nutrient broth) is generally used. Nutrient broth is a complex media, which is used to identify unknown microorganisms, the nutrient requirements of which are unknown, from the environment. This media contains all sources of nutrients in unknown chemical composition. Media generally consist of the following ingredients like peptone, beef extract, yeast extract, and NaCl. Peptone is a protein hydrolysate of soyameal, bean, milk protein like casein, etc. Protein serves both as the amino acid source and as nitrogen source for the microorganisms. Beef extract is an aqeous extract from lean beef and serves as organic and protein source. Beef extract also contain organic acids, carbohydrates, vitamins, etc. Yeast extract is an aqueous extract from Brewer's yeast. It is the best source of vitamin B12 (cyanocobalamin). NaCl is an electrolyte which acts as a pH stabilizer.

Materials Required

Conical flask

Test tube

Cotton

Measuring jar

Media

Nutrient Broth

Peptone	5 g
Beef extract	3 g
Yeast extract	2 g
NaCl	5 g
Distilled water	100 ml

pH 7–7.2

Procedure

Take a clean conical flask with 100 ml of distilled water. Weigh the above nutrients for 100 ml and dissolve them in distilled water. Adjust the pH range to 7– 7.2 and sterilize it at 121°C for 15 minutes for 15 lbs atmospheric pressure. Cool the media and pour 5 ml of media in

(a)   (b)

Liquid media (a) Control without inoculation
(b) Positive control turbid appearance.

each test tube under aseptic condition. Inoculate the microorganism with inoculation needle. Incubate the test tube at 37°C for 12 hours. Keep one test tube (nutrient broth) as a control without inoculation.

Experiment 2 Preparation of Solid Media (Nutrient Agar) and Cultivation of Micro-organism

This media is mainly used for the surface cultivation of microorganisms, for which the solidifying agent (agar) is used. Agar is a solidifying agent with boiling point 80–90°C and solidifying point 40–45°C. It is a sulphated polymer which contains D-galactose, 2,3 anhydrogalactose and glucouronic acid. It is extracted from red algae, *Rhodophyta* spp. It was first used by Hesse and his colleagues. It is used for the surface cultivation of microorganisms since it is not easily degraded by microorganisms.

Materials Required

Conical flaks

Petriplates

Cotton

Measuring jar

Media

Nutrient Agar

Peptone	5 g
Beef extract	3 g
Yeast extract	2 g
NaCl	5 g
Agar	20 g
Distilled water	1000 ml

pH 7–7.2

Procedure

Take a clean conical flask with required amount of distilled water. The above ingredients except agar are weighed accurately and dissolved in required amount of distilled water. Adjust the pH range to 7–7.2 and then add the solidifying agent, agar. Sterilize the flask at 121°C (or) 15 pounds atmospheric pressure for 15 minutes, cool and pour the media into the plates (15 ml) and allow to solidify. Inoculate the petri dish with microorganism using inoculation needle. Incubate the plates at 37°C for 12 hours. Keep one petri plate without inoculation, as a control.

Experiment 3   Preparation of Solid Media (Nutrient Agar Slants) and Cultivation of Microorganism

This media is formulated for the cultivation and subculturing of microorganisms.

Materials Required

Conical flask

Test tube

Cotton

Measuring jar

Media

Nutrient Agar

Peptone	5 g
Beef extract	3 g
Yeast extract	2 g
NaCl	5 g
Agar	20 g
Distilled water	1000 ml

pH 7–7.2

Procedure

Take a clean conical flask with required amount of distilled water. Weigh the above ingredients, except agar, accurately and dissolve in required amount of distilled water. Adjust the pH range to 7 – 7.2 and then add the solidifying agent (agar). Sterilize the media at 121°C for 15 minutes (or) 15 lbs. Cool the media and pour approximately 5 ml nutrient agar into the test tubes and keep in the slanting position aseptically. Allow them to solidify. Inoculate the slant with the

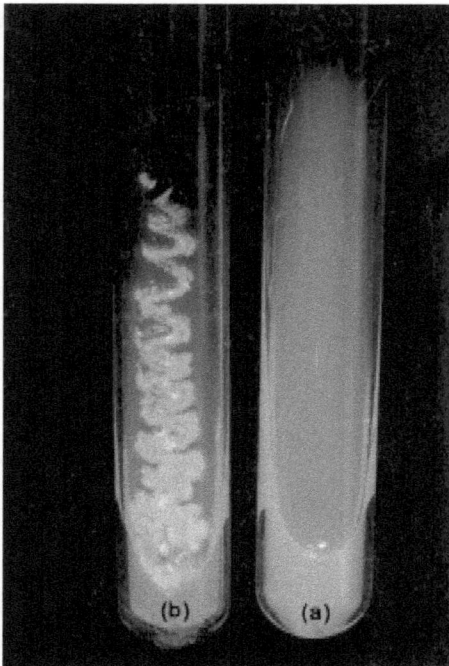

Solid media in tube nutrient slant (a) Control tube without Inoculation (b)
Growth of the organism in streaked area.

microorganisms using inoculation needle and incubate at 37°C for 12
hours. Keep one nutrient agar slant without inoculation as control.

Experiment 4   Observation of Various Microbial
Colonies on Plates

Microorganisms are ubiquitous in nature. This test is mainly to prove
the ubiquitous nature of microorganisms. The culture media contains
complex raw materials such as peptone, beef extract. Peptone serves
as the nitrogen source, beef extract serves as the source of
carbohydrates and organic nitrogen compounds. Yeast extract serves
as vitamin B source and as growth-promoting substance.

Materials Required

Petri plates
Measuring jar
Cotton
Conical flask

Media

Nutrient Agar

Peptone	5 g
Beef extract	3 g
Yeast extract	2 g
NaCl	5 g
Agar	20 g
Distilled water	1000 ml

pH 7–7.2

Procedure

Prepare the nutrient agar plates by using plating technique. Open the plates at various places for five minutes time interval. Close the

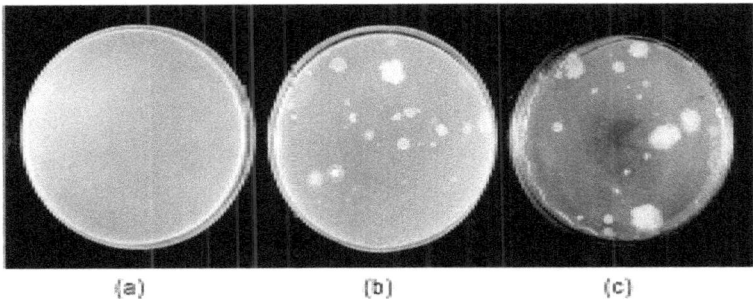

(a)   (b)   (c)

Photograph shows the ubiquitous nature of microorganism
(a) Control plate without exposure to environment
(b & c) Plates exposed to various environments.

plates and incubate at 37°C for 12 hours. Keep one petri plate (without opening) as control.

Experiment 5   Serial Dilution Technique

In nature, microbial populations do not segregate themselves by species but exist with a mixture of many other cell types. In the laboratory, these populations can be separated into pure cultures. The growth of the mass of the cell in the same species is called 'Pure Culture'. In the technical sense, a pure culture is one grown from a single cell. These pure cultures contain only one type of organism and are suitable for the study of their cultural, morphological, and biochemical properties.

Materials Required

Conical flask

Test tubes

Cotton

'L' rod

Measuring jar

Petri dish

Media

Nutrient Agar

Peptone	0.5 g
Beef extract	0.3 g
Yeast extract	0.2 g
NaCl	0.5 g
Agar	2 g/100 ml

pH 7–7.2

Procedure

- Prepare ten sterile test tubes with 4.5 ml sterile distilled water.
- Take 0.5 ml of given culture suspension using a sterile pipette and pour it into the first test tube.

- The dilution range of first test tube is 10^{-1}. Using a cyclomixer instrument mix the diluted culture suspension well.

- Then take 0.5 ml of culture suspension from the 10^{-1} dilution tube and pour it into the second tube. The dilution range of the second tube is 10^{-2}.

- Like this serially dilute the culture suspension up to 10^{-10} dilution.

  Note: For isolation of pure culture, the next step is spread plate technique.

Experiment 6　Spread Plate Technique

This technique requires the previously diluted mixture of microorganism. During inoculation, the cells are spread over the surface of a solid agar medium with a sterile 'L' shaped bent rod.

- Prepare the nutrient agar plates by using the plating techniques.

- From the serial diluted samples take 0.1 ml using a sterile pipette and pour into the petri plate.

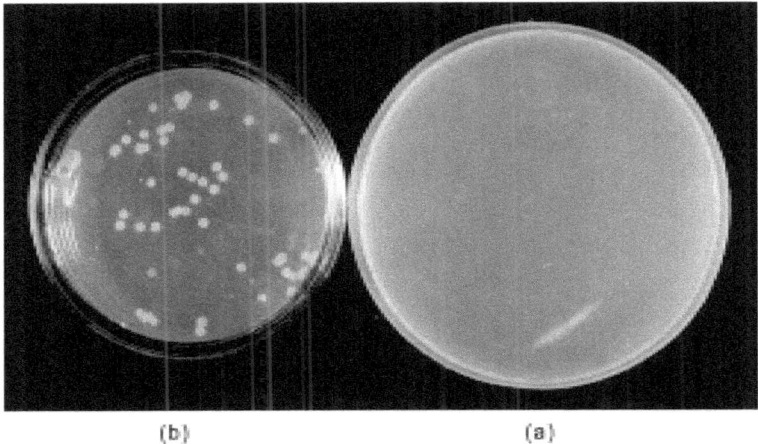

(b)　　　　　　　　　　　(a)

Spread plate technique
(a) Control plate without inoculum (b) Isolated colonies

- Dip the 'L' rod in alcohol for cleaning.
- After cleaning with alcohol, sterilize the 'L' rod in Bunsen flame.
- Spread the sample over the surface of the nutrient agar plates using 'L' rod.
- Likewise spread all the tube samples in separate petri plates.
- Mark each plate and incubate it at 37°C for 24 hours.
- Incubate a control plate without inoculation.
- Observe the result and count the colonies using colony counter.
- Record your values in a table.

Serial Number	Dilution range	No of colony forming units (CFU)	No of microbial cells
1	10^{-1}		CFU × 10 Cells
2	10^{-2}		
3	10^{-3}		
4	10^{-4}		
5	10^{-5}		
6	10^{-6}		
7	10^{-7}		
8	10^{-8}		
9	10^{-9}		
10	10^{-10}		

Experiment 7   Streaking Technique

The technique commonly used for isolation of discrete colonies initially require that the number of organisms in the inoculum be reduced. The resulting diminution of the population size ensures that following inoculation, individual cells will be sufficiently far apart on the surface of the agar medium to effect a separation of the different species present. Streak plate technique is a rapid qualitative isolation method. It is essentially a dilution technique that involves spreading a loopful of culture over the surface of an agar plate. Although many types of procedures are performed, the four way, or quadrant streak method is described.

Materials Required

Petri dish

Measuring jar

Cotton

Media

Nutrients Agar

(Composition of nutrient agar as given in previous experiments.)

Procedure

• Prepare the nutrient agar plates by plating technique.

• Streak one loopful of organisms (single colony) over the area near the edge of the plate.

• Flame the loop, cool it for 5 seconds and make 5 or 6 streaks from area one through area two.

• Flame the loop, again cool it and make six or seven streaks from area two through area three.

• Flame the loop again and make as many streaks as possible from area three to area 4.

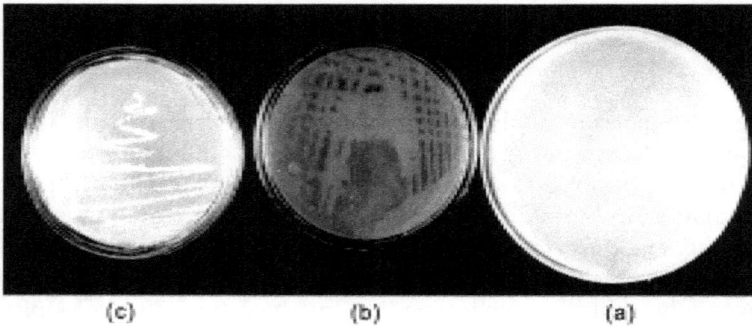

(c)   (b)   (a)

Plates shows the various streaking of microorganism (a) Control plate
without inoculum (b) Quadrant streaking (c) Simple streaking

- Flame the loop before putting down.
- Incubate the petri dish at 37°C for 24 hours. Keep one petri dish
  without streak as a control.

Staining

The different staining techniques include gram staining, simple
staining, capsule staining, negative staining and spore staining
techniques.

Experiment 8   Gram Staining

This technique is mainly performed to observe the nature of the cell
wall of the given microorganism.

The primary stain used in gram staining is crystal violet. It stains
all the bacterial cells purple violet. A mordant, Gram's iodine, acts as a
fixative that binds with the primary stain and forms insoluble Crystal
violet–iodine complex (CV–I Complex). Gram-positive bacterial cell
takes only crystal violet and Gram's iodine binds into negative
ribonucleic acid compound of the cell wall and forms the RNA–CVI
complex. Alcohol acts as lipid solvent and protein dehydrating agent.
In gram-positive cells the low lipid concentration is important and

the small amount of lipid is dissolved by the action of decolouring agent alcohol, which results in the formation of cell wall pores, then closed by the alcohol dehydrating effect. The tightly bound primary stain is difficult to remove and the cells remain violet in colour. In gram-negative cells, because of the high lipid concentration in their cell wall which is dissolved by the action of alcohol, large pores are created in the cell wall. Since these pores are large, they do not close by the dehydration of the wall proteins. So the unbound crystal violet complex is released. When the counterstain safranin is added, only the gram-negative cells take the counterstain. So they appear pink in colour.

Materials Required

Inoculation needle

Spiral lamp

Microscopes

Bacterial samples

Glass slides

Chemicals

Primary stain (Crystal violet)

Mordant (Gram's iodine)

Decolourising agent (Ethyl Alcohol 75%)

Counterstain (Safranin)

Procedure

- Smear the given samples on the clean glass slide.
- Dry the smear and heat-fix.
- Add crystal violet as a primary stain and keep for one minute. Wash with tap water.
- Add mordant as a fixative and keep for one minute. Wash with tap water.
- Add ethyl alcohol (95%) as a decolourising agent for 15–30 seconds and wash with tap water.

- Finally add the counterstain and keep for one minute. Wash with tap water.
- Blot the smear and dry with filter paper. Examine the slide under the microscope.

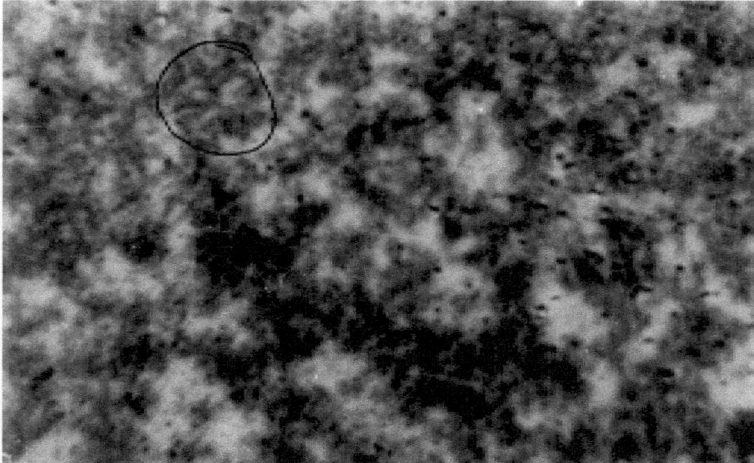

Photograph shows the gram staining of microorganism. Marked area indicates the rod-shaped gram-negative organism.

Experiment 9   Simple Staining

This technique is used for the identification of bacterial cell shape and cell arrangement.

A simple staining solution contains only one stain dissolved in a solvent and applied once to the bacteria. The purpose of simple stain is to colour the bacteria so that it may be easily seen. The stain is a positively charged chromogen. Since bacterial nucleic acid and certain cellular components carry negative charge, the negatively charged components attract the positively charged ions and bind tightly with each other. The basic stains are methylene blue (1–2 minutes) or carbol fuchsin (15–30 seconds).

Materials Required

Inoculation needle

Spirit lamp

Bacterial cultures

Slide

Chemicals

Methylene blue

Procedure

Preparation of smear

- Take a clean glass slide which is dust free. Incinerate the needle in the spirit lamp. Place a drop of distilled water on the slide.

- Transfer a bit of culture using incinerated needle into the glass slide.

- Make sure that only the top of the needle touches the culture to prevent the transfer of many cells.

- Spread the culture in an area of one inch.

- A good smear will appear as a thin whitish layer.

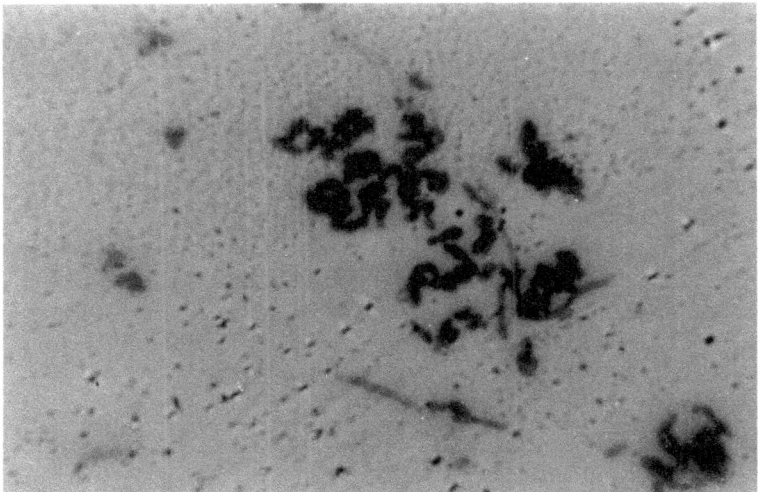

Photograph shows the simple staining of microorganism.

- Dry the smear and heat-fix.

  (Smear preparations prevent the bacterial cells from being washed off. During heat fixation bacterial proteins are coagulated and fixed on the slide. Heat fixation can be done by passing the glass slide with dried smear rapidly over the flame of the Bunsen burner 3 or 4 times.)

- Place the slide in the staining tray and flood the smear with methylene blue. Keep the stain for 2 minutes.

- Wash with water to remove excess stain.

- Dry the stain and observe the slide under the microscope.

Experiment 10   Capsule Staining

Capsule is a gelatinous outer layer that is secreted by the cell and that surrounds and adheres to the cell wall. It is made up of polysaccharide, a glycoprotein, or a polypeptide. Capsule staining is more difficult than other types of differential staining procedures because the capsules materials are water-soluble and may be dislodged and removed with vigorous washing. Smears should not be heated because the resultant cell shrinkage may create a clear area around the organism, and this is an artifact that can be mistaken for the capsule.

The capsule stain uses two reagents primary stain crystal violet. A violet stain is used to a non-heat fixed smear. At this point, the cell and the capsular material will take on the dark colour. Because of the nonionic charge of the capsule, the primary stain adheres to the capsule without binding to it. Since capsule is water-soluble, copper sulphate rather than water, is used to wash the purple primary stain out of the capsular material without removing the stain that bound to the cell wall. At the same time, it acts as a counterstain as it is absorbed into the decolourised capsule material. The capsule will now appear light blue in contrast to the deep purple colour of the cell.

Materials Required

Bacterial cultures

Bunsen burner

Inoculating loop or needle

Staining tray

Bitulous paper

Lens paper

Glass paper

Microscope

Chemicals

1% Crystal violet

20% copper sulphate $(CuSO_4 \cdot 5H_2O)$

Procedure

* Take a clean glass slide.
* Using sterile techniques prepare a heavy smear of each organism.
* Allow the smears to air-dry. Do not heat-fix.
* Flood the smear with crystal violet and let it stand for 5–7 minutes.
* Wash smears with 20% copper sulphate solution.
* Gently blot dry and examine under oil immersion microscope.

Experiment 11   Negative Staining

Negative staining requires the use of acidic stain such as India ink or nigrosin. The acidic stain, with its negatively charged chromogen, will not penetrate the cells because of the negative charge on the surface of bacteria. Thus, the unstained cells are easily differentiated against the coloured background.

The practical application of negative staining is twofold; First, since heat fixation is not required and the cells are not subjected to the distorting effects of chemicals and heat, their natural size and shape can be seen. Second it is possible to observe bacteria that are difficult to stain, such as some spirits.

Materials Required

Glass slide

Inoculation needle

Slant cultures

Bunsen burner

Chemicals

Nigrosin

Procedure

- Place a small drop of nigrosin on the top of the clean slide.
- Using sterile techniques, place a loopful of inoculum from the culture in the drop of nigrosin and mix.
- With the edge of a second slide held at a 30°C angle and placed in front of the bacterial suspension, push the mixture to form a thin smear.
- Air-dry. Do not heat-fix the slide.
- Repeat steps 1 to 4 for slide preparation of various cultures.
- Examine the slides under oil immersion.

Experiment 12   Spore Staining (Schalffer–Fulton Method)

This technique is mainly to perform the procedure for differentiation between bacterial spore and vegetative cell forms.

The members of the anaerobic genera *Clostridium* and *Desulfotomaculum* and the aerobic genus *Bacillus* are examples of organisms that have the capacity to exist either as metabolically active vegetative cells or as highly resistant metabolically inactive dormant cells called 'spores' during unfavorable environmental condition, i.e. particularly with the exhaustion of nutritional carbon source, these cells have the capacity to undergo sporogenesis and give rise to a new extracellular structure called the 'endospore' (for more details see first chapter).

In practice, the spore stain uses two different reagents. The primary malachite green, unlike most vegetative cell types that stain by common procedures, the spore because of its impervious coats, will not accept the primary stain easily. Heat is applied for further penetration. After the primary stain is applied and the smear is heated, both the vegetative cells and spores will appear green.

For mild decolourisation, tap water can be used. Once the spore accepts the malachite green it cannot be decolourised by tap water, which removes only the excess primary stain. The spore will remain green. On the other hand, the stain does not demonstrate a strong affinity for vegetative cell components; the water removes it and there cells will be colourless

Safranin is a counterstain, which is red in color and used as the second reagent to colour the decolorized vegetative cells, which will absorb the counterstair and appear red. The spores retain the green of the primary stain.

Materials Required

Bacterial cultures

Bunsen burner

Hot plate

Staining tray

Inoculating loop

Glass slides

Bitulous paper

Lens paper

Microscopes

Chemicals

Malachite green

Safranin

Procedure

- Take clean glass slides.
- Make individual smears of different bacterial cultures in the usual manner
- Using sterile techniques allow smear to air-dry, and heat-fix in the usual manner.
- Flood the smears with malachite green and place on a warm hot plate, allowing the preparation to smear for 2 to 3 minutes. Caution: Do not allow stain to evaporate; Replenish stain as needed.
- Prevent the stain from boiling by adjusting the hot plate temperature.
- Remove slides from the hot plate, cool and wash under running tap water.
- Blot dry with bitulous paper and examine under oil immersion microscope.

BIOCHEMICAL TESTS

Biochemical tests are mainly used to carry out for the characterization of Enterobacteriaceae. The metabolic nature, substrate utilization and nutrients requirements can be identify using biochemical test. The tests includes indole test, methyl red test, Voges Proskauer test, citrate test, are collectively denoted as IMViC test.

Experiment 13   Indole Test

This is mainly used to perform whether the given organism has the ability to utilize tryptophan or not.

When tryptophan is supplemented in the medium as peptone it is degraded and converted into indole, pyruvic acid and ammonia. This reaction is carried out by the help of the enzyme tryptophanase. The ability of few organisms to produce indole may be used as one of differentiating characters for Entrobacteriaceae family. The presence of indole in the culture medium can be detected by calorimetric reaction or by the help of the chromogen Kovac's reagent.

Tryptophan $\xrightarrow{\text{Tryptophanase}}$ Indole + pyruvicacid + ammonia

Kovac's Reagent + Indole $\longrightarrow$ cherry red colour

Materials Required

Conical flask

Test tube

Measuring jar

Cotton

Inoculation needle

Media

Peptone broth

Peptone	20 g
NaCl	5 g
Distilled water	1000 ml

pH 7.4

Reagents

Kovac's reagent

Amyl alcohol	150 g
Para dimethyl amino benzaldehyde	10 g
Concentrated hydrochloric acid	50 ml

Dissolve the aldehyde in the alcohol and cool it. Then add acid and store in the refrigerator.

Procedure

Prepare the peptone broth as given in the composition. Sterilize the medium at 121°C for 15 minutes. After sterilization, inoculate

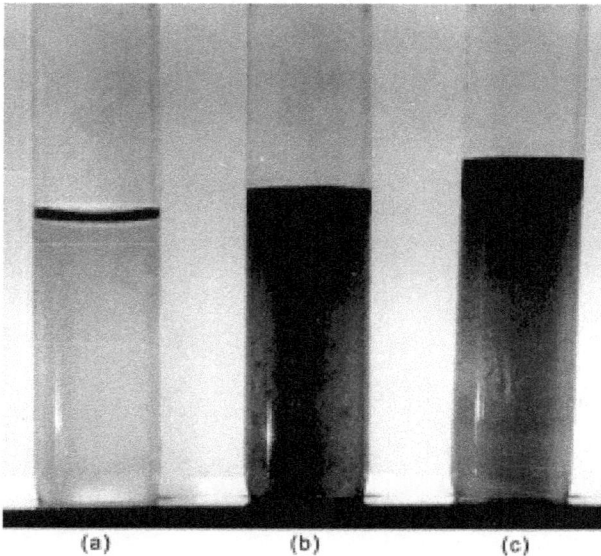

(a)　　　　　　(b)　　　　　　(c)

Photograph shows the indole test indicator Kovac's
reagent (a) Control tube without inoculation
(b) Positive control (c) Negative control.

the tubes with *E. coli* and *Klebsiella*. Incubate the tubes at 37°C for 24
hours. After incubation, add 0.2 ml of Kovac's reagents for each 5 ml of
the culture medium. Observe the result.

Experiment 14　　Methyl Red Test

Methyl red test is used to differentiate the gram-negative intestinal
bacteria on the basis of end products formed by the fermentation of
glucose.

Genera of bacteria such as *Escherichia, Salmonella, Proteus,* and
Aeromonas, ferment glucose to produce lactic acid, acetic acid, succinic
acid and formic acid, CO_2, H_2O, and ethanol. Accumulation of these
acids lowers the pH of the medium to 5.0 or less. If methyl red is added
to such a culture, the indicator turns red, which indicates that the
organisms are greatly gas producers too. In addition to that the

organisms produce the enzyme formic hydrogenase which splits the formic acid into equal parts of CO_2 and H_2O.

Indicator

Methyl red is an acid–base indicator. In the presence of acid the indicator turns cherry red in colour. Subsequently in the presence of base it turns yellow in colour.

Materials Required

Conical flask

Test tube

Cotton

Inoculation needle

Media

Glucose	5 g	
Peptone	5 g	
KH_2PO_4	5 g	
Distilled water	1000 ml	
pH 7.0–7.2.		

Reagents

Methyl Red

Procedure

Prepare MR-VP broth and sterilize the media at 121°C for 15 minutes Pour the media into test tubes. Inoculate the culture in all the tubes except only one tube to incubate as control. Incubate the tubes at 37°C for 24 hours. After incubation add 3–4 drops of methyl red indicator and observe the result.

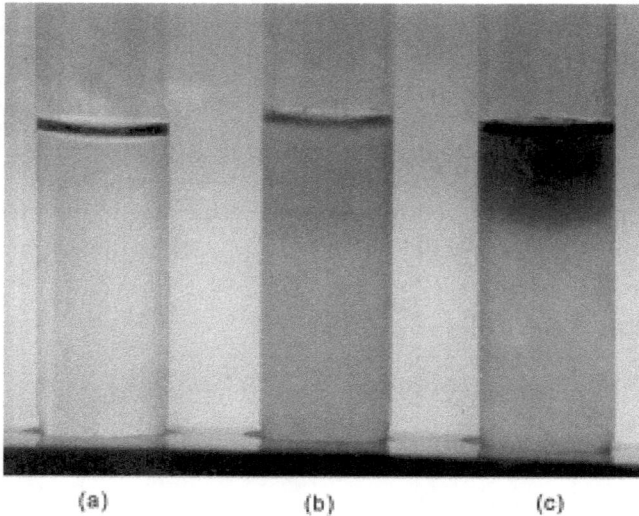

Photograph shows Methyl red test indicator Methyl red (a) Control tube without inoculation (b) Negative control (c) Positive control.

Experiment 15   Voges Proskauer Test

This test is mainly performed to find out whether the given organism produces non-acid end product such as alcohols.

Some species of bacteria (*Bacillus, Erwinia, Areomonas*) produce lot of 2, 3 butanediol and ethanol. Instead of acids unfortunately there is no satisfactory test for 2, 3 butanediol. However acetoine (acetyl methyl carbinol), a precursor of 2, 3 butanediol can be easily detected with Barritt's reagent. While adding barritt's reagent the medium changes to pink or red in the presence of acetoine. This is Voges Proskauer test.

Materials Required

Conical flask

Test tubes

Culture

Media

MR-VP Broth

Glucose	5 g
Peptone	5 g
KH_2PO_4	5 g
Distilled water	1000 ml

pH 7–7.2

Reagents

Barritt's Reagent

Solution A   6 gms of α-napthol in 100 ml of 95% ethyl alcohol.

Solution B   Dissolve 16 gms of potassium hydroxide in 100 ml of water.

Mix solution A and solution B.

Procedure

• Prepare MR-VP broth and sterilize the medium at 121°C for 15 minutes. Pour the medium in test tubes. Inoculate the culture in all the tubes and keep only one tube without inoculation as control.

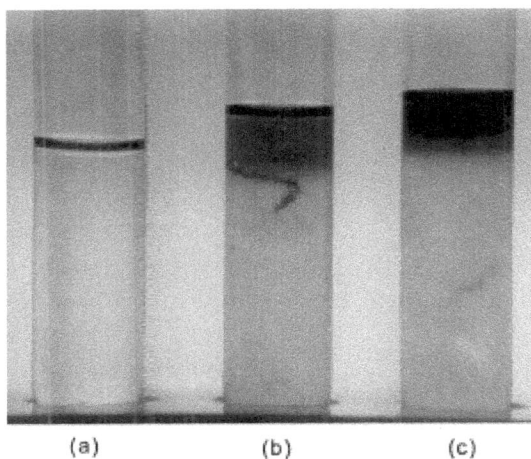

Photograph shows the Voges Proskauer test (a) Control tube without inoculation (b) Negative control (c) Positive control.

- Incubate the tube at 37°C for 24 hours. After incubation add small amount of Barritt's reagents and observe the result.

Experiment 16   Citrate Test

This test is mainly carried out to differentiate enteric organisms on the basis of their ability to ferment citrate as a sole source of carbon.

In the absence of glucose or lactose. Some microorganisms use citrate as the carbon source which depends on the enzyme citrate permease. The enzyme citrate permease acts upon the substrate citrate which converts to oxaloacetic acid and acetate. These are then enzymatically converted to pyruvic acid and CO_2. The medium is supplemented with sodium citrate as the carbon source. During the reaction the organism utilize citrate and leaves sodium in the medium. The medium becomes alkaline as the CO_2 combines with sodium and water, which forms sodium carbonate, an alkaline product. The sodium carbonate changes the bromothymol blue from green to deep Prussian blue colour.

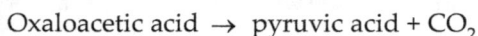

Citrate $\xrightarrow{\text{Citrase}}$ oxaloacetic acid + acetic acid

Oxaloacetic acid → pyruvic acid + CO_2

CO_2 + $2Na^+$ + H_2O → Na_2CO_3 sodium bicarbonate (Prussian blue colour).

Bromothymol blue + Sodium bicarbonate $\xrightarrow{\text{Alkaline pH}}$ Green to Prussian blue colour.

Indicator

The indicator is an acid–base indicator which appears green colour in acidic condition and turns Prussian blue in colour in basic condition.

Materials Required

Test tubes

Measuring jar

Conical flask

Inoculation needle

Bunsen burner

Media

Simmon Citrate Agar

Ammonium hydrogen phosphate	1 g
Dipotassium phosphate	1 g
Sodium chloride	5 g
Sodium citrate	0.2 g
Agar	15 g
Bromothymol blue	0.08 g/ml

pH 6.8 ± 0.2

Procedure

Prepare the simmon citrate agar slant tubes. Inoculate one set of tubes with *E. coli* and the other with *Klebsiella*. Incubate the tube at 37°C for 24 hours. After incubation observe the colour change.

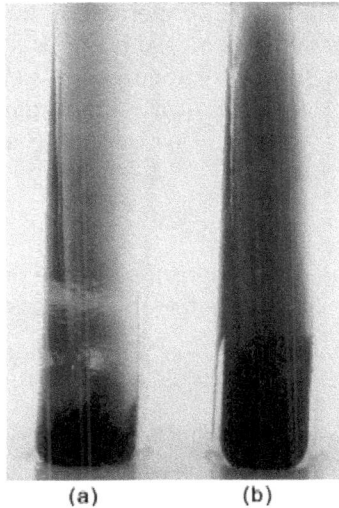

(a)   (b)

Photograph shows the citrate utilization (a) Control tube without inoculation (b) Positive control, indicator Bromothymol blue

Experiment 17 Triple Sugar–Iron Test

This test is mainly performed to differentiate between the Enterobacteriaceae family and also to determine the following:

i.   Sugar fermentation

ii. Gas production

iii. H_2S production

This is used to differentiate the group Enterobacteriaceae according to their ability to ferment lactose, sucrose, glucose and production of H_2S. The fermentation reaction of sugar will help to distinguish Entero bacteriaceae from other gram-negative intestinal bacilli. TSI slant contains 1% each of lactose, sucrose and glucose in a concentration of 0.1%. The phenol red indicator is also incorporated in the medium to detect the carbohydrate fermentation.

Sugar Fermentation

Acid butt, alkaline slant (yellow butt, red slant) indicates the glucose fermentation but not lactose fermentation and sucrose *fermentation*. Alkaline butt and alkaline slant (red butt, red slant) indicates none of glucose, sucrose or lactose fermentation. Acid butt, alkaline slant (yellow butt, yellow slant) indicates the fermentation of lactose and sucrose.

Gas Production

The production of gas can be identified by the presence of bubbles. The high production of gas breaks the agar media or pushes it upwards.

H_2S Production

H_2S production can be identified by a blackening of the butt. This is due to the reaction of H_2S with ferrous ammonium sulphate supplemented in the medium, which forms ferrous sulphide.

Indicator

Phenol red is an acid–base indicator, present in the medium. Phenol red gives yellow colour in acidic condition, while it turns red in colour in alkaline condition.

Materials Required

Test tubes

Conical flask

Inoculation needle

Measuring jar

Bunsen burner

Media

TSI Agar

Peptone	20 g
Lactose	10 g
Sucrose	10 g
Glucose	1 g
Sodium chloride	5 g
Ferrous ammonium sulphate	0.2 g
Phenol red	0.025 g
Agar	15 g/1000 ml

Procedure

Prepare the TSI agar slant such that it has a butt and slant. Inoculate the first set of tubes with *E. coli* and inoculate the second set of tubes with *Salmonella* and third set tubes with *Shigella* and last set of tubes with *Proteus*. Stab the culture in butt once and streak in the slant. Incubate the tubes at 37°C for 24 hours.

(a)   (b)   (c)   (d)

Photograph shows the triple sugar–iron test (a) Control without
inoculation (b) Tube with acid butt, acid slant (c) Tube with H₂S
production (d) Tube with acid butt, alkaline slant. Indicator phenol red.

Experiment 18   Carbohydrate Fermentation

This test is mainly performed to test how the given culture of organisms
ferment different carbohydrates and produce gas.

The ability of an organism to attack and break down various
carbohydrates can be determined by the use of a suitable nutrient
medium containing the carbohydrate and an acid–base indicator. A
liquid medium is usually dispensed in test tubes "D–urham's tubes"
to collect some of the gas produced. Acid production is indicated by
change in colour. The formation of acid or gas is an indication that the
carbohydrate is utilized.

Materials Required

Test tube

Durham's tubes

Conical flask

Cotton

Media

Glucose broth

Lactose broth

Mannitol broth

Sucrose broth

Arabinose broth

Inulin broth

Maltose broth

Media Composition

Respective sugars in respective broth	=	1 g (in the case of inulin, 1.25 g)
Peptone	=	5 g
Beef extract	=	3 g
Bromothymol blue	=	1.2 ml
Distilled water	=	1000 ml

pH = 7.2 ± 0.2

Procedure

Prepare all the broths and sterilize the broth. Inoculate the given organisms. Simultaneously prepare the sub-culture of the given organism. Before inoculation, put the Durham's tubes in the inverted position into each tube without any bubble. Incubate the tubes at 37°C for 24 hours and observe the result.

(a)   (b)   (c)

Photograph shows the carbohydrates utilization (a) Control tube
without inoculation (b) Tube with gas production (positive)
(c) Tube without gas production (negative).

Biochemical charateristics of various microorganisms

S. No	Organism	I	M	Vi	C
1	*E. coli*	+	+	–	–
2	*Citrobacter*	d	+	–	–
3	*Salmonella*	–	+	–	+
4	*Shigella*	d	+	–	+
5	*Klebsiella*	–	–	–	–
6	*Enterobacter*	–	–	+	+
7	*Proteus*	d	+	–	+
8	*Bacillus*	+	–	+	+
9	*Pseudomonas*	–	–	–	+

10	*Staphylococcus*	–	+	+	–
11	*Clostridium*	–	+	–	–
12	*Vibrio cholerae*	+	–	–	–
13	*Brucella*	–	–	–	–

I   Indole test
M   Methyl red test
Vi Voges Proskauer
C   Citrate test
D   Results different in different strains/species
+   Positive result
–   Negative result

Study of Microbial Enzymatic Activity

Experiment 19   Starch Hydrolysis

This test determines mainly the utilization of starch with the help of extracellular enzymes produced by bacteria.

Starch is a linear polymer of glucose molecules linked together by glycosidic bond. Starch as such cannot be transported into the cell because of its high molecular weight to assimilate starch for energy catabolic reactions. It must be degraded into basic glucose units by starch-splitting enzymes. The enzymes are secreted by the microorganism into the medium that degrades starch primarily to glucose. The resulting low molecular weight soluble glucose are now able to pass into the cell for energy production through glucose.

Material Required

Conical flask
Petri dishes

Measuring jar

Inoculation needle

Media

Starch Agar

Peptone	5 g
Beef extract	3 g
Soluble starch	2 g
Agar	15 g
Distilled water	1000 ml

pH 7.0

Reagents

Iodine solution

In the presence of starch iodine gives blue colouration.

Procedure

Take a clean conical flask with 100 ml of distilled water. Weigh accurately the above ingredients except agar and dissolve in 100 ml of

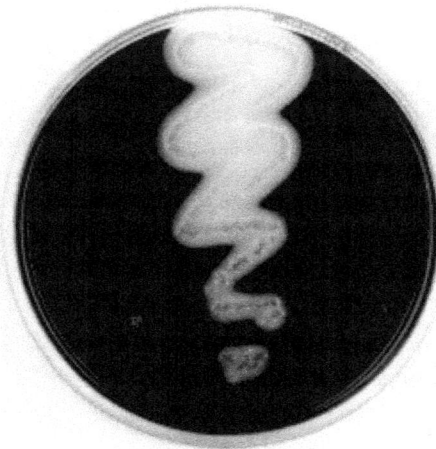

Photograph shows the hydrolysis of starch. Clear white zone around the colonies indicates the starch utilization.

distilled water. Adjust the pH range to 7.0–7.2 and then add solidifying agent agar. Sterilize the media and pour at 121°C for 15 minutes. Cool the media and pour it into the sterilized petri dishes in aseptic condition and allow to solidify. After the solidification, streak one part of the plates with *Bacillus* species and the other plates with *E. coli* and incubate at 37°C for 12 hrs. After incubation, flood the plate with iodine solution and observe the colour change.

Experiment 20　Protein Hydrolysis

This is mainly performed to check the ability of microorganisms to secrete the proteolytic enzymes that are capable of degrading the protein, casein.

Being a major milk protein, casein is a macromolecule made up of amino acids linked by peptide bonds. Molecules of this magnitude are not able to pass through the cell membrane. The protein should undergo stepwise degradation into peptones, polypeptones, dipeptides and finally to amino acids for cellular nutrition. This is mediated by cellular exoenzymes called 'Proteases' and this process is called peptidization or proteolysis.

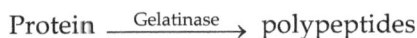

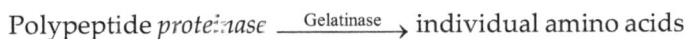

$$\text{Protein} \xrightarrow{\text{Gelatinase}} \text{polypeptides}$$

$$\text{Polypeptide } proteinase \xrightarrow{\text{Gelatinase}} \text{individual amino acids}$$

Materials Required

Petri plates

Conical flask

Measuring jar

Cotton

Bunsen burner

Media

Skimmed Milk Agar

Skimmed milk powder　100 g

Peptone   5 g

Agar   15 g

Distilled water   1000 ml

pH 7.2

Procedure

Prepare skimmed milk agar plates and inoculate with test culture as a quadrant streak and incubate the plates at 37°C for 24 to 48 hours. After incubation observe the results.

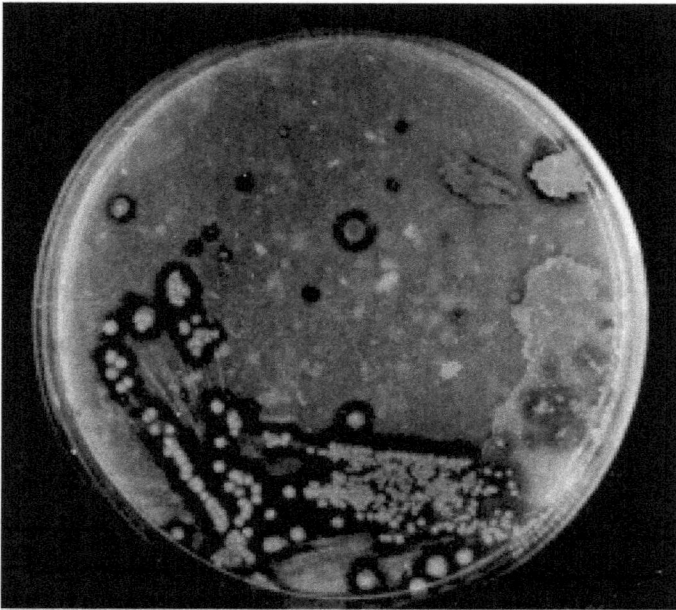

Photograph shows the utilization of milk protein
indicated by clear zone around the colonies.

Experiment 21   Lipid Hydrolysis

This test is used to test the ability of the given organism to utilize lipids by producing the enzymes.

A lipase enzyme is produced by lipid-utilizing bacteria. Due to the action of this enzyme the lipids are converted into glycerol and fatty acids

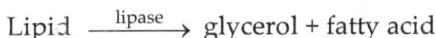

$$\text{Lipid} \xrightarrow{\text{lipase}} \text{glycerol} + \text{fatty acid}$$

Fatty acids + copper sulphate $\longrightarrow$ bluish green colour.

Materials Required

Conical flask

Petri dish

Measuring jar

Media

Nutrient agar with 0.3 ml of vegetable fat (vanaspathi)

Peptone	5 g
Beef extract	3 g
Yeast extract	2 g
NaCl	5 g
Distilled water	1000 ml

pH 7–7.2

Procedure

Emulsify the fat (0.3 ml of vegetable fat) in the nutrient agar by shaking it thoroughly just before pouring into plate. Inoculate the plate with *E. coli* and *Pseudomonas* and incubate at 37°C for seven days. After incubation flood the agar with a saturated aqueous solution of copper sulphate and keep it for 10–15 minutes. Pour off the excess reagents from the plate and observe the result.

Experiment 22   Catalase Test

This test is used to determine the ability of some microorganisms to degrade hydrogen peroxide by producing the enzyme catalase. During aerobic respiration, microorganisms produced hydrogen peroxide and in some cases, an extremely toxic superoxide. Accumulations of these substances results in death of the organism unless they can be enzymatically degraded. These substance are produced when aerobes, facultative anaerobic and micro aerophiles use the aerobic respiratory pathway, in which oxygen is the final electron acceptor during degradation of carbohydrates for energy production. Organisms capable of producing catalase rapidly degrade hydrogen peroxide

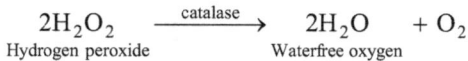

$$2H_2O_2 \xrightarrow{\text{catalase}} 2H_2O + O_2$$
Hydrogen peroxide   Waterfree oxygen

In the case of catalase-negative organisms, the enzyme super oxide dismutase degrades the toxic superoxides. The inability of strict anaerobes to synthesize catalase, peroxidase, or superoxide dismutase may explain why oxygen is poisonous to these microorganisms. In the absence of these enzymes, the toxic H_2O_2 cannot be degraded when these organism are cultivated in the presence of oxygen. Catalase production can be determined by adding the substrate H_2O_2 to an appropriately incubated trypticase soy agar slant culture. If catalase is present, the chemical reaction mentioned is indicated by bubbles of free oxygen gas (O_2). This is a positive catalase test; the absence of bubble formation is a negative catalase test.

Materials Required

Broth cultures of various organisms

Bunsen burner

Inoculating loop

Glassware

Marking pencil

Chemicals

3% hydrogen peroxide

Procedure

• Using sterile techniques, inoculate each experimental organism into its appropriately labelled tube by means of a streak inoculation. The last tube will serve as a contol.

- Incubate all cultures for 24–48 hours at 37°C.
- After incubation allow 3–4 drops of the 3% hydrogen peroxide to flow over the entire sugar of each plate or slant tube.
- Examine each culture for the presence or absence of bubbling or foaming. Record your results in the chart.

Experiment 23   Oxidase Test

This test is performed on the basis of cytochrome oxidase activity. Oxidase enzymes play a vital role in the operation of the electron transport system during aerobic respiration. Cytochrome oxidase catalyses the oxidation of reduced cytochrome by molecular oxygen (O_2) resulting in the formation of H_2O or H_2O_2. Aerobic bacteria as well as some facultative anaerobic and microaerophiles exhibit oxidase activity. The oxidase test aids in the differentiation among members of the genera *Neisseria* and *Pseudomonas* which are oxidase-positive and the Enterobacteriaceae which are oxidase-negative.

The ability of bacteria to produce cytochrome oxidase can be determined by the addition of the test reagent, p-amino dimethyl aniline oxalate on the colonies. The light pink reagent serves as an artificial substrate, donating electrons and thereby becoming oxidized to a blackish compound in the presence of the oxidase-free oxygen. Following the addition of test reagents, the development of pink, then maroon, and finally black colouration on the surface of the colonies is the indication of cytochrome oxidase production and represents a positive test. No colour change or a light pink colouration on the colonies is indicative of the absence of oxidase activity, that is a negative test.

Materials Required

24–48 hours of positive and negative cultures in plates

Bunsen burner

Inoculating loop

Glassware

Marker

Chemicals

P - aminodimethyl aniline oxalate

Procedure

*   Take the 24–48 hours incubated culture plates.

*   Add two or three drops of the P-amino dimethyl aniline oxalate to the surface of the growth of each test organism.

*   Observe the growth for the presence or absence of colour changes from pink, to maroon and finally to purple. Positive test (+) is a colour change in 10–30 seconds. Negative test (–) is no colour change or light pink colour. Record the results.

Experiment 24   Measurement of Cell Size

This is the technique to measure the bacterial cell using micrometer.

Some microorganisms can be seen only under the light microscopes. A suitable scale for the measurement of cell size for these microorganisms is provided by the ocular micrometer. The ocular micrometer has a scale divided into 100 divisions of arbitrary length and there are markings like 10, 20 and so on. The exact distance between two lines (one division) of the ocular micrometer will depends on the magnifying power of the ocular and the objective lens used in combination. The distance will vary with the objective. On the other hand, the stage micrometer has a definite scale where 1 mm is divided into 100 divisions and each is a length of 10 microns. Five consecutive divisions are marked by slightly longer lines indicated by still longer graduation. Ten consecutive small divisions are indicated by still longer graduations.

Materials Required

Microscope

Ocular micrometer

Stage micrometer

Smeared glass slide

Procedure

- Remove the eyepiece and insert the ocular micrometer on the circular hole (disc-shaped diaphragm).
- Replace the eyepiece and observe. The ocular micrometer can be observed by sharp focus.
- Mount the stage micrometer on the microscope stage and using the low power (10X) objective. Bring into clear focus the graduation on the micrometer scale.
- By rotating the ocular, superimpose images of the ocular micrometer scale and stage micrometer scale. See the places where the two scales are coinciding.
- Mark the final member of ocular division that corresponds to the known distance in the stage micrometer.
- Calculate the value of one division of ocular micrometer for different objectives and replace the stage micrometer with the slide bearing the microorganism.

Calculation

Under 45X Objective

5 divisions of ocular micrometer = 15 divisions of stage micrometer

$$= 15 \times 10 \ \mu m$$

∴ 5 divisions of ocular micrometer = 150 μm of stage micrometer

$$1 \text{ division of ocular micrometer} = \frac{150}{5} = 30 \ \mu m$$

1 division of ocular micrometer = 30 μm

$$= 1 \times 30 = 30 \ \mu m$$

The total length of the bacterial cell under low-power objective lens covers one division

So the length of the bacterial cell is = 30 μm.

Experiment 25   Growth Curve of Bacteria

This technique is generally used to study the growth characteristics of the given bacterial culture by plotting a growth curve. The graph will give the information regarding the generation time of a bacterial culture. This technique mainly requires the inoculation of viable cells into a sterile broth medium and incubation of the culture under optimum temperature, pH and oxygen concentration. Under these conditions, the cells will reproduce rapidly and the dynamics of the microbial growth can be characterized by means of a population growth curve, which is constructed by plotting the increase in cell numbers versus time of incubation. The curve can be used to delineate stages of the growth cycle. It also facilitates measurement of cell numbers and the rate of growth of a particular organism under "standardized time", the time required for a microbial population to double. A typical growth has four stages (i) Lag phase (ii) Log Phase (iii) Stationary phase (iv) Decline phase. (details of these stages are given in chapter 6).

This experiment requires the aliquots of a 24 hours shake flask culture be measured for population size at intervals during the incubation period. The curve will be plotted on semilog paper by using two values for the measurement of growth. Growth curve can be plotted by two methods. (i) The direct method requires enumeration of viable cells in serially diluted samples of the test cultures taken at 30-minute intervals. (ii) The indirect method uses spectrophotometric measurement of the developing turbidity at the same 30-minute intervals.

You can determine generation time with indirect and direct methods by using data on the growth curve. In direct determination can be done by simple exploration from the log phase as illustrated in the figure. Selected two points on the optical density scale such as 0.2 and 0.4 that represent a doubling of turbidity. Using a ruler, extrapolate by drawing a line between each of the selected optical densities on the ordinate and the plotted line of the growth curve. Then draw perpendicular lines from these end points on the plotted lines of the growth curve to their respective time intervals. With this information, determine the generation time as follows.

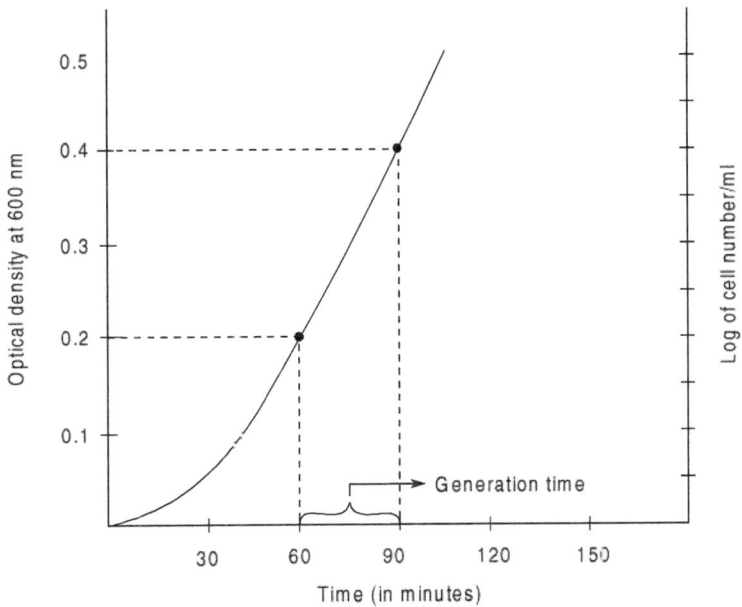

$$GT = t_{(OD\ 0.4)} - t_{(OD\ 0.2)}$$

GT = 90 minutes − 60 minutes = 30 minutes

The direct method uses the log of cell number scale on the growth curve and the following formula

$$GT = \frac{t \log_2}{\log b - \log B}$$

where,

GT = generation time

B = number of bacterial cells at some point during the log phase.

b = number of bacterial cells at a second point of the log phase

t = time in hours or minutes between B and b

Materials Required

10–12 hour (log phase) culture

37°C water bath shaker incubator

Spectrometer

Cuvettes

Colony counter

Sterile petri dish

Sterile pipette

Glass marker

Beaker

Bunsen burner

Sterile water blanks

Media

Nutrient Agar

Procedure

- Take 99 ml sterile water blanks into six sets of three water blanks each. Label each set of blank as to time of inoculation (t_0, t_{30}, t_{60}, t_{120}, t_{150}) and the dilution to be effected in each water blank (10^{-2}, 10^{-4}, 10^{-6}).

- Label six sets of four petri dishes as to time of inoculation and dilution to be plated (10^{-5}, 10^{-6}, 10^{-7}).

- Liquify the four bottles of nutrient agar in a water bath. Cool and maintain at 45°C.

- Using a sterile pipette, add 5 ml of *E. coli* log phase culture into the flask containing 100 ml of nutrient broth.

- The initial OD (t_0) should be 0.08–0.1 at 600 nm.

- After the t_0 OD, shake the culture flask. Aseptically transfer 1 ml to the 99 ml water blank labelled t_0 10^{-2} and continue to dilute serially to 10^{-4} and 10^{-6}.

- Place the culture flask in water bath shaker. Set at 120 rpm at 37°C and time for the required 30-minute interval.

- Shake the t_0 dilution bottle, plate the t_0 dilution on the appropriate labelled t_0 plate. Aseptically pour 15 ml of the molten agar into each plate and mix by gentle rotation.

- Likewise at each 30-minute interval, shake the flask and aseptically transfer 5 ml into cuvette. Take the OD values. Simultaneously take 1 ml of culture from the flask and transfer it to 10^{-2} water blank. Serially dilute the sample and plate it in the petri dish.

- When the pour plate culture hardens, incubate them in an inverted position for 24 hours at 37°C.

Staining Reagents

Capsule Stain

Crystal violet (1%)

Crystal violet (85% dye content)	1.0 g
Distilled water	100 ml

Copper sulphate solution (20%)

Copper Sulphate ($CuSO_4.5H_2O$)	20.0 g
Distilled water	80 ml

Gram Stain

Crystal violet

Solution A

Crystal violet (90% dye content)	2.0 g
Ethyl alcohol (95%)	20 ml

Solution B

Ammonium oxalate	0.8 g
Distilled water	80.0 ml

Gram's iodine

Iodine	1.0 g
Potassium iodide	2.0 g
Distilled water	300 ml

Ethyl alcohol 95%

Ethyl alcohol (100%)	95 ml
Distilled water	5 ml

Safranin

Safranin O	0.25 ml
Ethyl alcohol (95%)	10 ml
Distilled water	100 ml

Nigrosin

Nigrosin, water-soluble	10.0 g
Distilled water	100 ml

Note: Immerse in boiling water bath for 30 minutes.

Formalin	0.5 ml

Note: Filter twice through double filter paper.

Spore Stain

Malachite green

Malachite green	5 g
Distilled water	100 ml

Safranin   Same as given in Gram stain.

GLOSSARY

Anabolism The utilization of energy generated by catabolism for the biosynthesis of cell components.

Aerobe or oxygenic Refers to the organism which reduces oxygen and hence said to be aerobe.

Anaerobe Organism which does not require oxygen for its growth.

Anaplerotic pathway Ancillary pathway of some metabolism where each step is mediated by separate enzymes.

Amphibolic pathway The central pathway which participates both in energy-yielding and energy- requiring processes simultaneously.

Autotrophs Microorganisms which use CO_2 as their sole or principle source of carbon.

Antiport system A type of transport system in which the transported substances move in opposite direction, i.e., one substance moves in the net inward direction while the second moves in the net outward direction.

Apoenzyme Simple enzymes which are made up of protein part only.

Ammonification The production of ammonia from the aminoacids by oxidative deamination.

Aerobic respiration Energy-yielding processes in which the electron transport chain acceptor is an oxidized organic molecule other than oxygen.

Balanced growth Increase of cell components in proportion to each other.

Bioluminescence The emission of light from the electron transport chain of certain living organisms. The enzyme luciferase picks up electrons from flavoproteins in the electron transport chain and emits some of the electron energy as photons of light.

Catabolism   Generation of chemical energy suitable for the mechanical and chemical processes of the cells.

Chemoheterotrophs   Organisms that degrade organic material and use organic carbon to produce the building blocks.

Chemotaxis   The movement of a bacterium towards chemical substances.

Chemotrophs   Organism that obtains energy from the oxidation of chemical compounds (either organic or inorganic).

Chemoorganotrophs   Organisms that use organic compounds as source of energy, hydrogen, electrons and carbon for biosynthesis.

Chemolithotrophs   Organisms that oxidize reduced inorganic compounds such as iron, nitrogen or sulphur molecules to derive both electron and energy for biosynthesis.

Cryptic growth   The balanced events of death and growth in a stationary phase.

Continuous culture   Maintaining microbes continually in the logarithm phase.

Cardinal temperature   Refers to the maximum, optimum and minimum temperatures.

Chemical energy   The energy released when organic or inorganic compounds are oxidized.

Carbon dioxide fixation   A process by which inorganic carbon dioxide becomes incorporated (fixed) into the structure of organic compounds within the cell.

Diaminopimelic acid   Amino acid present in N-acetyl glucosamine of peptidoglycan, and is not found in any other protein.

De novo pathway   The synthesis of nucleotides by forming a parent molecule, IMP.

Deamination   Removal of the amino group from amino acid.

Denitrification   The process of transformation of nitrogen to gaseous nitrogen by microorganisms in a series of biochemical reactions.

Exergonic   The liberation of chemical-free energy during the chemical reaction.

Endergonic (Energy-consuming process) process which requires an input of free energy and denotes the total endothermic energy requirement including heat.

Extremophiles Microorganisms, which live in extreme environments like high salt, high pressure, temperature, etc.

Eurythermal Denotes a wide range of temperature of an organism.

Enzymes Protein biological catalysts which have great specificity for the reaction catalysed and the molecules acted on.

Fermentation Denotes the breakdown of complex molecules into simple compounds with the help of an enzyme.

Free energy The energy released that is available to do useful work. The change in free energy during a reaction is expressed as $\Delta G^{\circ'}$.

Growth factors Organic nutrients that are precursors of some components and cannot be synthesized by the organism.

Growth Final expression of the physiology of an organism.

Generation time Time required for a single cell to divide.

Glycolysis The process where a molecule of glucose is degraded in a series of enzyme-catalysed reactions to yield 2 molecules of pyruvate.

Gluconeogenesis Defined as the biosynthesis of glucose from noncarbohydrate compounds.

Hypotonic solution A solution in which the external osmotic pressure is lower than the internal system.

Heterotrophs Organisms which use reduced organic compounds as source of both carbon and energy.

Holoenzyme The enzyme which is made up of protein and a small molecule of nonprotein part.

Homolactic fermentation The process of fermentation where the end product formed is lactic acid.

Heterolactic fermentation The production of lactic acid and ethanol, carbondioxide are as additional products through fermentation.

Isotonic medium The medium that has no osmotic pressure.

Isoenzymes The enzyme which exists in many physically distinct forms and all can catalyse the same reaction.

Kilocalorie   The quantity of heat energy necessary to raise the temperature of one kilogram of water by 1°C. 1 kilogram is equal to 4.184 KJ.

Lipoproteins   Lipids covalently linked to proteins.

Lipopolysaccharides   Large complex molecules that contain both lipid and carbohydrate which consists of three parts.

Lithotrophs   Organism uses reduced inorganic substances as their electron donor.

Lag phase   The initial phase of growth where the division and metabolism takes place slowly in order to synthesize the enzyme and other cell constituents for the uptake of a new medium.

Metabolism   Assembly of biochemical reactions which are employed by the organism for the synthesis of cell materials and for the utilization of energy from their environment.

Magnetotaxis   Movement of flagella towards the earth's magnetic field.

Minor elements   The elements which can exist in the cell as cations and play a variety of roles.

Mixotrophs   Organisms that combine chemolithotrophic and heterotrophic metabolic processes.

Macromolecules   Proteins, carbohydrates, amino acids, nucleic acids and complex lipids which are cell constituents.

Mineralization   The conversion of complex organic compounds into simple inorganic compounds or into their constituent elements by the help of microorganisms.

Methanogenesis   The production of methane by enzymatic reaction of anaerobic methanogenic bacteria.

Nutrients   The substances used in biosynthesis and energy production and required for microbial growth.

Nucleoside   The attachment of pyrimidine bases or purine bases through nitrogen atoms to a pentose sugar.

Nucleotides   The attachment of nucleoside with a phosphate group.

Nitrogen cycle   The sequence of changes in the conversion of atmospheric nitrogen to simple organic compounds and complex organic compounds in the tissue of plants, and animals, and the eventual release of the nitrogen back to the atmosphere.

Nitrification The conversion of ammonia to nitrate by microorganisms.

Nonsymbiotic microorganisms Free-living microorganims like *Azotobacter, Klebsiella,* and *Cyanobacteria* that fix atmospheric nitrogen, the process called as nonsymbiosis.

Nitrogen fixation The reduction of molecular atmospheric nitrogen into ammonia.

Organotrophs Organisms that extract electron or hydrogen from organic compounds.

Oxidation Removal of electron or electrons from a substance.

Pleomorphic Microorganisms that have the ability to change shape.

Phototaxis The movement of flagella towards light.

Peptidoglycan Component of cell wall composed of enormous polymers of many identical structure. Peptidoglycan is otherwise called as murein.

Protoplast The cell membrane and the internal organelles.

Phototrophs Organisms which trap light energy during photosynthesis.

Photoautotrophs Microorganisms that use light energy and CO_2 as their carbon source.

Photoorganotrophic heterotrophs Most of this group are photosynthetic and they use organic matter as their electron donor.

pH The negative logarithm of hydrogen ion concentration.

Poikilothermic Organisms whose temperature varies with that of the external environment.

Protein synthesis Synthesis of protein using the genetic information in a messenger RNA as a template.

Post-translational modification The processing of protein by forming hydrogen bonds and van der Waals, ionic and hydrophobic interactions after the protein synthesis.

Pasteur effect   The slower rate of glucose utilization by a microorganism which is growing aerobically by respiratory mechanism than the same organism growing anaerobically.

Respiration   Any process in which there is transfer of electrons and release of energy, whether terminating at molecular oxygen or not.

Reduction   Denotes the addition of electrons to a substance. A hydrogen atom (H) consists of an electron plus a proton. When the electron is removed, the hydrogen atom becomes proton (H^+).

Spore   A special resistant dormant structure of a cell under unfavourable condition.

Sporulation   The process of formation of endospore within a vegetative cell.

Spheroplast   Spherical structure of an organism which surrounds the cellular contents, plasma membrane and remaining outer wall layer.

'Shift down' change   If the present environment is nutritionally less abundant than the previous one, the change is referred as shift down change.

'Shift up' change   When the present environment of an organism is more nutritionally abundant than the previous one, the change is known as shift up change.

Stenothermal   Organisms that show a small range of growth temperature.

Signal sequences   Short sequences of amino acid residues which direct a protein in its appropriate location in the cell.

Salvage pathway   The utilization of purines and pyrimidines in the form of free bases as well as nucleotides, when these compounds are supplied in the medium.

Saturated fatty acids   Fatty acids containing only a single bond.

Symport   The transport of two materials occurring in the same direction is referred to as symport.

Simple diffusion   A transport mechanism, where there is no utilization of energy and where protein is not mediated.

Sulphur assimilation Reduction of sulphur from sulphur source is termed as assimilatory sulphate reduction.

Taxis The measurement of a bacterium towards or away from a particular stimulus by its flagella.

Teichoic acid Component of gram-positive cell wall consisting primarily of an alcohol (such as glycerol and ribitol) and phosphate. There are two classes of teichoic acids—lipoteichoic acid and wall teichoic acid.

Translocation The movement of mRNA to a distance equal to three bases long from A site to P site.

Unbalanced growth The type of growth where the components of a new cell increase in a non-constant relationship to each other.

Unsaturated fatty acids Fatty acids containing double bond e.g linolenic acid.

Uniport The protein-mediated transport occurring in a single direction.

SUGGESTED READINGS

Alcamo, I Edward. *Fundamentals of Microbiology*. California: Benjamin/ Cummings Publishing Co. Inc., 1994.

Benson, Harold J. *Microbiological Applications: A Laboratory Manual in General Microbiology*. London: Irwin, 1995.

Caldwell, Daniel R. *Microbial Physiology and Metabolism*. Boston: McGraw-Hill, 1995.

Cappuccino, James G and Natalie Sherman,. *Microbiology: A Laboratory Manual*. New York: Addison-Wesley Publishing Co., 1999.

Collins, C H, Patricia M Lyne, and J M Grange,. *Microbiological Methods*. New Delhi: Reed Elsevier India Private Limited, 2001.

Funke, Berdell R. *Study Guide for Microbiology: An Introduction*. California: Benjamin/Cummings Publishing Co. Inc., 1998.

Heritage, J, E G V Evans, and A Killington,. *Introductory Microbiology*. Cambridge: Cambridge University Press, 1996.

Inglis, T J J. *Microbiology*. Edinburg: Churchill Livingston, 1997.

Ingraham, John L and Catherine A Ingraham,. *Introduction to Microbiology*. Singapore: Thomson Asia Private Limited, 2000.

Johnson, Ted R and Christine L Case,. *Laboratory Experiments in Microbiology*. California: Benjamin/Cummings Publishing Co. Inc., 1995.

Lim, Daniel V. *Microbiology*. New York: Wadsworth Publishing Company, 1989.

Madigan, Michael T, John M Martinko, and Jack Parker. *Biology of Microorganisms*. New York: Parker Publishing Company Inc, 1997.

Mckane, Larry and Judy Kandel,. *Microbiology: Essentials and Applications*. New Delhi: Tata McGraw-Hill Publishing Co. Ltd., 1996.

Moat, Albert G, John W Foster, and Micheal P Spector,. *Microbial Physiology*. New York: John Wiley & Sons, 1995.

Murray, Robert K, Dary I K Granner, and Peter A Mayes,. *Harper's Illustrated Biochemistry*. New Delhi: Tata McGraw-Hill Publishing Co. Ltd., 2003.

Pelczar, Michael J, E C S Chan, and Noel R Krieg,. *Microbiology*. New Delhi: Tata McGraw-Hill Publishing Co. Ltd., 1996.

Plummer, David T. *Introduction to Practical Biochemistry*. New Delhi: Tata McGraw-Hill Publishing Co. Ltd., 2003.

Prescott, Lansing M, John P Harley, and Donald A Klein. *Microbiology*. Boston: McGraw-Hill, 1999.

Robertis, E D P De and E M F De Robertis,. *Cell and Molecular Biology*. New Delhi: B.I. Waverly Pvt. Ltd, 1998.

Sawhney, S K and Randhir Singh,. *Introductory Practical Biochemistry*. New Delhi: Narosa Publishing House, 2002.

Schlegel, Hans G. *General Microbiology*. Cambridge: Cambridge University Press, 1996.

Stanier, Roger Y, John L Ingraham, and Mark L Wheelis. *General Microbiology*. New Delhi: Macmillan India Limited, 1995.

Starr, Secie and Ralph Taggart,. *Cell Biology and Genetics*. Belmolt: Wadsworth Publishing Company, 1998.

Tilton, Richard C. *Microbiology: Pretest, Self-Assessment and Review*. New Delhi: Tata McGraw-Hill Publishing Co. Ltd., 1996.

Tortora, Gerard J, Berdell R Funke, and Christine L Case,. *Microbiology: An Introduction*. California: Benjamin/Cummings Publishing Co. Inc., 1994.

INDEX